THE PARTICLE HIERARCHY PARADIGM

Charles "Don" Briddell

Using loop analytics of Tometry
(topological loop fractal skew geometry) and
the Sierpinski Triangle Fractal diagram to quantize
how loops of action interacting three-dimensionally
generate the known mass and energy values of particles.

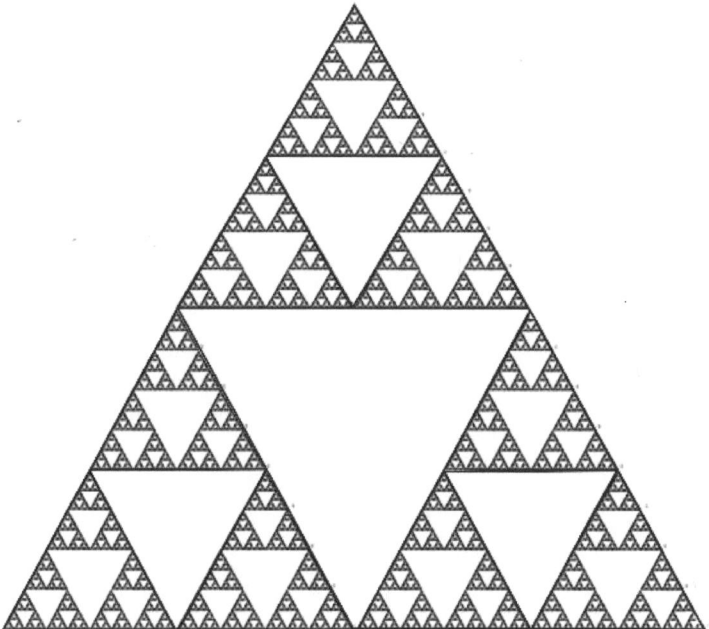

Sierpinski Triangle Fractal

The Particle Hierarchy Paradigm

Published by:

Field Structure Institute
Don Briddell
8002-A Dollyhyde Rd.
Mount Airy, MD 21771

donbriddell@fieldstructureinstitute.org
www.FieldStructureInstitute.org

ISBN 978-1-7357041-0-4 (paperback)
ISBN 978-1-7357041-1-1 (epub)
ISBN 978-1-7357041-2-8 (mobi)

69 illustrations, photos, and drawings by Don Briddell except as noted.

Computer graphics and contributing photographs by Joseph Clinton

ACKNOWLEDGEMENTS

Stanley Wysocki, (b. July 1943 - d. April 2020) structuralist, visual mathematician, retired Industrial Design and Architecture instructor at Pratt Institute, Brooklyn, NY, who was my college roommate and with whom I co-authored a senior thesis titled "Form and Structure Theory" at Pratt in 1966. This thesis project lead to Tometry and Field Structure Theory. Stanley was a rare genius and an important early contributor to what became Field Structure Theory.

William Katavolus, professor of Industrial Design at Pratt Institute, Brooklyn, NY originally inspired Stanley Wysocki and me to study Form and Structure Theory, which led us to structural physics.

Roger Tobie, a master's student at Pratt Institute Brooklyn, NY when I first met him and a friend thereafter. As a result of Roger's encouragement, the department chairperson allowed Stanley Wysocki and me to dispense with some other studies so as to give us more time to work on our Form and Structure thesis.

My wife, Victoria Moo, who from the time we first met (1965) has encouraged the work, and added to it in subtle and profound ways, the most significant of which was pointing me in the direction where I could find answers, namely to Indian philosophy.

Swami Chidananda, Mother Rema Veeramani, and Swami Krishnananda of India, who each in their own way helped immensely in my formation of ideas about field consciousness and its relationship to the physics of fields. These greats of Indian metaphysics were crucial. It was in Dec. 1999, a few days before the new millennium in Lonavala, India on a retreat with Swami Chidananda that I discovered the significance of the Sierpinski Triangle Fractal to the loop architecture of particles.

Joseph Clinton, a design scientist, educator, geometer, inventor, structural historian, and a student of Buckminster Fuller. Among many other things, he is credited with discovering the Clinton–Fieldstructure. Today Joe Clinton is considered a foremost spokesperson and authority on Design Science, i.e., synergetics, geodesics, and tensegrity. Working with him has been an ongoing source of amazement and revelation.

Domina Spencer, PhD, fifty-year Professor emeritus in mathematics and author of numerous articles and books on mathematics and electrodynamics at University of Connecticut, Storrs, CT., who after I had given a presentation of Field Structure Theory, indelibly inspired me with one sentence when she got up and said, "That is the best lecture I have heard in the last fifty years!"

Carl Littmann, a structuralist whose geometry has revealed profound basic relationships. He informed me, some years ago that the particle I needed to consider was the lambda hyperon, for which I am deeply appreciative.

CONTENTS

Foreword ix

Preface xi

CHAPTER 1 How Loops and Twistings Quantify Mass and Energy

Values 1

1 - Loop Particle Hierarchy 1

2 - Summary. 16

3 - Loop Families 23

4 - Analyzing the Sierpinski Triangle Fractal 25

5 - Using the Neutrino as the Unit of Measurement 29

6 - The Proton as a Loop Structure 35

7 - The Question of Symmetry 43

8 - Fractional Values of Particles 48

9 - Particle Hierarchy 49

10 - Neutrino 51

11 - Energy Spectrum of First-Generation Particles 54

12 - Electron's Relationship to Atomic Structure 55

13 - Summary of the Fieldstructure Hierarchy 61

14 - Basic Attributes of Matter in Terms of

Loops and Twists 65

15 - Real-Matter (RM) and Anti-Matter (AM) 66

16 - The Neutron 68

17 - Loop Condensement...69

18 - Summary 74

CHAPTER 2 **Loop Architecture of Particles**........................**77**

19 - Loop Architecture of Particles 77

20 - Assumptions..79

21 - Tometry and Field Structure Theory (FST)..............82

22 - Loops – The Energy-Centric View of Reality..........82

23 - Fieldstructures...97

24 - Space 99

25 - Energy as Loop Circuits................................102

26 - The *Structor* – The Fieldstructure Form Taken by
    Particles...106

27 - The *SuperStructor* – The Fieldstructure Form
    Taken by Atoms..120

28 - The *Mala* – The Fieldstructure Form
    Taken by Molecules 124

29 - Fields 126

Terms And Definitions 133

Bibliography..143

FOREWORD

An introduction to "structural physics".

Joseph D. Clinton 02/27/2020

Don Briddell is introducing us to a new way of doing physics. Physics thus far has been unable to describe the structure of the universe. "We have descriptions but not explanations." Similar to the thoughts of Stephen Wolfram, the mathematician, Benoit Mandelbrot, has helped us realize this universe has fractal architecture. Fractals simply change structure and can be iterated throughout scale. A successful theory of how the universe works will have to include fractal architecture. Coupled with loop dynamics, the form and structure of the natural world becomes apparent and quantifiable.

Nature does not separate problem solving into disciplines. It integrates all elements into whole systems. We must remember mathematics is a language that can only help explain abstract thought and physical reality. The transformation and unity of all things begin with connections and relationships between the physical and metaphysical duality of Universe. All that is physical is energy. Metaphysics is the instructions for the behaviour of the physical. As the physical and metaphysical make connections, their relationships initiate transformations of energy to take on forms that can be measured. The structure of a system is dependent upon interaction of localized energy fields within and externally acting upon it.

This book will introduce the reader to "structural physics", as a new way of doing physics. It explains a way to model the microscale structure with macroscale models made possible because the universe is governed by fractal architecture that can be scaled up and down with the structural hierarchy found in Field Structure Theory.

PREFACE

"It is a rather unfortunate fact (for mathematics) that much of the creative introduction of new geometric ideas is done by non-mathematicians, who encounter geometric problems in the course of their professional activities. Not finding the solution in the mathematical literature, and often not finding even a sympathetic ear among mathematicians, they proceed to develop their solutions as best they can and publish their results in the journals of their own disciplines."   Branko Grunbaum (1981)

Lord Kelvin and Peter Tait at the turn of the 20th century, gave up on working with macroscale materials to see if knots on a string could reveal something about the structure of the elements. Since then efforts to make macroscale models of particle and atom structures have been abandoned. With the advent of Quantum Mechanics, the only analytic used by physics has been mathematics. Begun in the nineteen-fifties, a cadre of enthusiasts began considering the world of form and structure and began developing a new analytic that came to be known as *Design Science*. Buckminister Fuller, Kenneth Snelson, William Katavolus, Arthur Loeb, Joseph Clinton, Jay Kappraff, Haresh Lalvani, John Bell, Stanley Wysocki, Roger Tobie, and I among others, took up structural studies in the fifties and sixties.

Structuralists are design scientists who study structure as a stand-alone science in its own right. Structuralists build macroscale models, mostly three-dimensional, that look for insights about the natural order of our world. The realization is that macroscale modeling of forms and structure have relevance to structure at all scales of reality.

The purpose of this book is to explain why and how structure analytics modeled in the macroscale can and has revealed the architecture of fundamental mass and energy.

Relying on math to understand fundamental structure has revealed much about nature, but in the end has proved problematic. Mathematics led those doing fundamental physics down a rabbit hole and into probabilities, chance and random behavior, a world foreign to our macroworld. Most physicists today believe that fundamental reality can only be known by the use of arcane probability mathematics. Determinism is dead as far as physics is concerned. What is worse, the physics community insist their predicament of seeing the world as governed by chance and probabilities is the only view allowed.

Einstein questioned the practice of allowing mathematics to lead before there was structural understanding. Using mathematics, physics has concluded that nature at its fundamental level is indeterminate, that it's governed by chance. Determinism once held as sacred in the Industrial Age became blurred in the Quantum Age to such an extent that physics abandoned it. The only way the math being used could bring order to the world of atoms and particles was to understand reality in terms of probabilities. With probabilities, nature is understood in generalities and determinism became essentially crowd control; what individuals (atoms and particles) in the crowd were up to remained a mystery.

Physics reasoned that since discrete notions about the microscale structure were not understandable with an analytic imported from macroscale experience, implied that microscale and cosmic scale of reality did not share some kind of unity. The math they developed for all things small and all things large was incompatible. **Not noticed was the fact that what was going on in the macroscale has not been properly understood.** This realization was a major shock to me. It meant mathematicians and society for that matter was holding wrong assumptions about the nature of reality. **To question the major underlying assumptions of one's view of reality is another shock even greater than the one just mentioned.** To surmount basic assumptions is truly challenging.

What got missed at the macroscale was that the math developed for physics was based on the behavior of objects in a field, not on the behavior of a field itself. Western culture never even suspected the reality of energy fields until Michael Faraday demonstrated them as a reality around 1850. Still today what I call *field consciousness* has not been assimilated into the Western mode of consciousness. Westerners dwell on objects in a field, not the field in the object. It is the Eastern modes of consciousness that specialize in field consciousness. I have gone into this issue extensively in a larger book to follow this book and will not dwell on it further here. It is sufficing to say this issue exists and needs attention.

Science began with the study of objects, not with the study of fields. It encountered fields not through the efforts of the mathematician, but rather through the efforts of an experimenter, Michael Faraday, working with macroscale materials. The existence of fields was not the invention of mathematicians. That came later once Faraday experimentally discovered them.

Fields can only be seen by implication. To the naked eye they are invisible. They are revealed by how they affect matter and this is what Faraday did. Now days, fields are fully respected for their primary role in the affairs of the objective world, but until the discovery of *fieldstructures*, there has not been a way to model a field empirically. With Field Structure Theory, we can actually *see* how fields operate.

Physics has neither been able to explain why particles have mass, nor has physics shown how those values came to be. This is because objects (particles) are derived from field energy. The problem has been that physics has assumed objects create fields. While fields and objects are inseparable, it is a mistake, as will be seen here, to give objects (particles) a causal status. If you can't explain how particles get their mass and energy, the universe will remain a mystery. We have descriptions, but not explanations. It is time for an explanation!

To address this problem, this book will introduce **"structural physics"**, a new way of doing physics. The fieldstructures made with material objects that you will see in this book, link fractal number patterns to patterns found in nature. Structural physics is interested in the mechanics of how action forms fields. Structural physics has found that structures properly built in the macro-scale utilize the same structural system nature uses throughout the universe. Energy in the form of electromagnetic loops and the effect those energy loops have on the Plenum[1] from which they are derived, reveal the relationship between field energy and particle mass.

In his book entitled "Six Easy Pieces", Richard Feynman, a Nobel Laureate in physics said in regards to a discussion on the gravity force, *"All we [physicists] have done is to describe* how *the earth moves around the sun, but we have not said* what makes *it go. Newton made no hypothesis about this; he was satisfied to find what it did without getting into the machinery of it.* No one has since given any machinery...The great laws of mechanics are quantitative mathematical laws for which no machinery is

---

1   *Plenum is the name used in this discourse for the substratum of reality. See "terms" at the back of this book for a definition.*

available. *Why we use mathematics to describe nature without a mechanism behind it, no one knows. We have to keep going because we find out more that way."*

The "machinery," as Feynman calls it, is the issue. Field Structure Theory (FST) will reveal this machinery and apply Tometry to understand the mass and energy values that have been experimentally found by physics. Tometry is a new analytic designed to replace geometry. By following the old math and forcing it to adhere to a flawed understanding of structure, physics has ended up with three theories, (the microscale [quantum mechanics], the macroscale [classical mechanics], and the cosmic [General Relativity]. These theories do not relate to one another. The reason physics has not succeeded in producing a successful Unified Field Theory, or what has been pompously called *The Theory of Everything,* is due to the lack of understanding structure. Physics has wonderfully weighed, measured, and timed the fundamental aspects of nature and given us descriptions any theory of structure will have to collaborate. The explanation of how the universe works as a unified whole is still wanting.

It is now the *structuralist's* turn to have a crack at it. **The problem is how does nature determine the mass and energies of particles?** We know what the values are, but not how they are determined theoretically. Physics found out these values mostly through experiment, not through a far-reaching theory. It is now time to address not what the numbers are, that we know, but how they come to be.

Structural physics, as delineated here, has constructed a particle hierarchy. Unable to penetrate the wall of mystery that seems to surround reality, structural physics has returned to natural forms and structures to do fundamental research. Unlike physics, which sees the problem in terms of particle objects, field theory sees the problem in terms of energy fields; particles being condensed fields of energy. Radiant energy is deployed energy; deployed from condensed fields. A proton is a condensed field. Destroy a proton, as can be done by contact with an anti-proton, and the condensed fields of both the proton and anti-proton convert into radiant gamma rays revealing that all form is energy whether bound or unbound. To understand energy is to understand fields. It is imperative to have a structural understanding of how to model a field, structurally.

Structuralists have gone back to model making within the protective domain of nature to see what went wrong with the quantum particle approach taken by physics. To do this a method had to be found that would allow the structural physicist to model a field, preferably at the macroscale so what is happening can be observed. This idea that there was such a thing as a macroscale model of microscale reality has long been discounted as foolishness by physics, once Lord Kelvin and Peter Tait gave up on tying knots on string at the beginning of the last century.

By only working with natural forces and materials, structural research is prevented from introducing the unnatural to satisfy conjecture. If we are to know how the universe works, we have to develop an analytic to interpret our observations. Western science chose mathematics as the tool to probe fundamental physical reality, in part because modeling with macroscale materials was seen as ineffective. FST saw that the problem was not with materials as such, but with the structural system science was assuming to be correct. Physics could not figure out how to model fields with materials so they quit trying. From the FST perspective, science failed to realize how fields extend to everything at all scales and materials. The structural system capable of modeling fields at the macroscale with materials awaited the discovery of *Tometry*, nature's analytic.

After spending a hundred years, science is beginning to see it's not the objects in the field, but the field in the objects that will get us somewhere. The core of the problem, as I see it, has been that physics has been asking the wrong analytic to solve how the universe was thought to work, when the question of structure had not been properly understood to begin with.

Stanley Wysocki and I at Pratt Institute in 1965 began working with paper, wire, plastic tubing, wooden dowel rods and rope; hardly the tools of the trade in physics. We found these materials entirely capable of modeling basic structure that was not limited by scale. Almost immediately we began discovering structures previously unknown, as far as we could tell. It never ceased to amaze us how there could be families of form and structure yet undiscovered. The prevailing opinion in 1965 science was that a theory of form and structure could not be consummated using macroscale forms and structures. Buckminster Fuller and Kenneth Snelson were belittled for trying. The macroscale world was too different from what was being encountered in the microscale and cosmic scale to have any relevance. So, we became foolish and tried it anyway. Almost immediately we discovered fieldstructures, a new family of knotted loop three-dimensional geometric forms. To this day I am amazed that this family of form had not been previously discovered. As can be seen Figures 1.2 and 1.3, a fieldstructure is, in its simplest form three bent loops that are knotted together to create a nuclear polyhedron. Whoever thought loops could create polyhedra? It was astounding to us that not only could the tetrahedron be created with loops but all polyhedra could be made with loops. Stanley and I began to see that the enormity of forms in the universe did have something simple behind them. That something was the loop. It turns out the universe of forms, impossibly vast in number, springs from the interaction of something as simple as the loop. We immediately speculated that the loops that string theorists were talking about in physics were the same loops we were working with in our studio. It was at that moment we realized we were doing physics but with a different analytic.

Our approach was to model with material objects, knowing full well, that macroscale modeling had been deemed useless by physics. However, we felt that as long as we were working with natural forms and structures, we were safe from wandering outside of what is natural. Natural structures have a way of self-proofing. If they are not structural, they collapse. That safety is something mathematical physics does not enjoy. Math can describe just about everything, including that which does not exist in the natural world. Math can be very misleading, as for instance, with the Greek Ptolemy, who thought he had successfully mathematically proven that the earth was the center of the universe. His math was correct, but it wasn't describing our universe.

Stanley and I had a strong belief that nature was a structural continuum; that what went on in the macroscale world of our studio, was inseparably linked to what went on at the small and large extremes of natural world. Nature has mathematical structure, to be sure, but before the math can be applied correctly, the mathematician needs to know what it is that he or she wishes to describe.

It has been necessary in the course of developing Field Structure Theory to invent words when there were none to express the concept being described. When the new word occurs, it will be italicized and listed at the back of the book under "Terms and Definitions".

The analytic that Field Structure Theory will be using is called "Tometry", short for <u>Topological Loop Fractal Skew Geometry</u>. The physical world is given to us as a *fait accompli*; and yet we see room for improvement. Our reality is in excellent working order. Still, however, we see that our position in the world can be improved by tweaking the mechanism. There is this sneaking suspicion held by humanity that nature, because it will respond to our wishes, has given us an open invitation to do so. Nature is not hiding anything, but it does insist we learn the rules and then obey them. At this time in our history, we are busy learning the rules. Harmonious and beneficial physics, the kind that works within the ecology of the natural system, is not going to begin until we learn the rules.

Structuralists believe that within our macro experience of the world is all we need to discover how the whole shebang works. At every scale of form and structure, be it physical, mental, or metaphysical, nature is presenting us, it would seem, with opportunity. Humanity has come to believe that nature has provided a staircase and invites us to climb to the stars.

Certain that nature wished humanity to prosper, the modern world responded. The rise from complacency to mastery began in the Renaissance. Without a handbook of instructions, we entered nature in hopes of unlocking its potentials. The question became, "What are we missing? How does nature work within its obvious unity?" Something went wrong in the process of knowing. With billions in treasure, billions of people time, and four-hundred years of effort, how it works is yet to be known. Something went wrong because physics ended up with the afore-mentioned three ideas about reality; in which the structure of microscale atoms and particles has nothing to do with macroscale models we may make on the kitchen table. Given our common beliefs about structure, I, along with everyone else, have been agreeing that nature is divided; that little things don't have any structural relationship to big things. Our *caught-in-the-middle* scale of experience doesn't look like either the little things or the big things. Everything physics has modeled at our macroscale is irrelevant to all things small and big. They insist models made with tubes, wires, and sticks can tell us nothing about the world of microscale physics where probabilities alone can tell us what is rational. As Stanley and I worked on our Form and Structure thesis at Pratt Institute, we began to suspect the problem was not with nature. It was with us. Nature was not complying with the analytics being used by physics. We needed to pursue this quest differently.

Since the early days of modern physics up to the time of Einstein (b:1879 - d:1953) physics made great progress. It started the century with proving the atom existed and ended the century with the identifying the particles that make up atoms. Since the nineteen-fifties, new methods of analysis and new discoveries have arrived. Fields, fractals, cellular automata, neutrinos, quarks, high-energy colliders, topologies, loop string theory, skew geometry, and a whole slew of new resonance particles among other things have become known. The thing that didn't arrive was a mathematics that adhered to nature's imperative, which is:

A geometry where the allowable forms correspond to the forms allowed by nature and none other.

The mathematicians, Benoit Mandelbrot and Stephen Wolfram have helped us realize this universe has fractal architecture. Fractals change everything. It means all structure at the macroscale can be iterated down (*downerated*) in scale or scaled up (*uperated*). All that was needed to weld fractal architecture to natural reality was the discovery of the loop's ability to generate structure, the structural system nature was using. Fractals implied that what could be made on the kitchen table, if they are natural forms, could also be the structural model for molecules, atoms and particles and for that matter, stars, galaxies and the whole universe, simply by iterating up or down the structural pattern and understanding how the form changes with inputs or outputs of energy.

What is needed is the correct structural system. For Stanley and I, that structural system became known as Field Structure Theory using the Tometry analytic. Having a way to model fields was the key to it all. Loops of action became the simple configuration that could create and model fields. These loop action fields could then model quantitatively energy and energy could then model mass. When it finally dawned on me how this is possible, a rush of tears welled up. I was overcome. Science had become a source of joy and happiness, the same joy and awe that art has been for me.

Lastly, several illustrations are repeated in this book. I have done this so the reader will not have to go searching for the image I had discussed elsewhere.

A note about the model on the cover of the book:
The two structures in color are Stereographs. By crossing your eyes, an image will appear in the middle between the two images right and left. Stare at the middle image until your mind reforms the image into a three-dimensional image. With a little practice it fuses into a single 3-D image quickly. The visual experience of seeing it in three-dimensions is so complete you can move your attention around the form and inspect it carefully.

CHAPTER 1

HOW LOOPS AND TWISTINGS QUANTIFY MASS AND ENERGY VALUES.

1 - Loop Particle Hierarchy

Field Structure Theory using Tometry (topological skew fractal geometry) and the Sierpinski Triangle Fractal show how loops of action interacting three-dimensionally are able to quantify mass and energy values. Importantly, Tometry permits all scales of reality to be understood with macroscale modeling. There is not a structure for small things (microscale) and another structure for large things (cosmic scale). Structure is a continuum of platform hierarchies. In nature the hierarchy of loops can be diagramed with the Sierpinski Triangle Fractal (STF).

The Sierpinski Triangle Fractal (STF) – each iteration starting with the first iteration is multiplied by three to form the next iteration.

Consider each line in the Sierpinski Triangle Fractal (STF) to be a loop. These are not Euclidian loops that have no dimension (thickness). These loops have dimensionality. Thus, we live in a

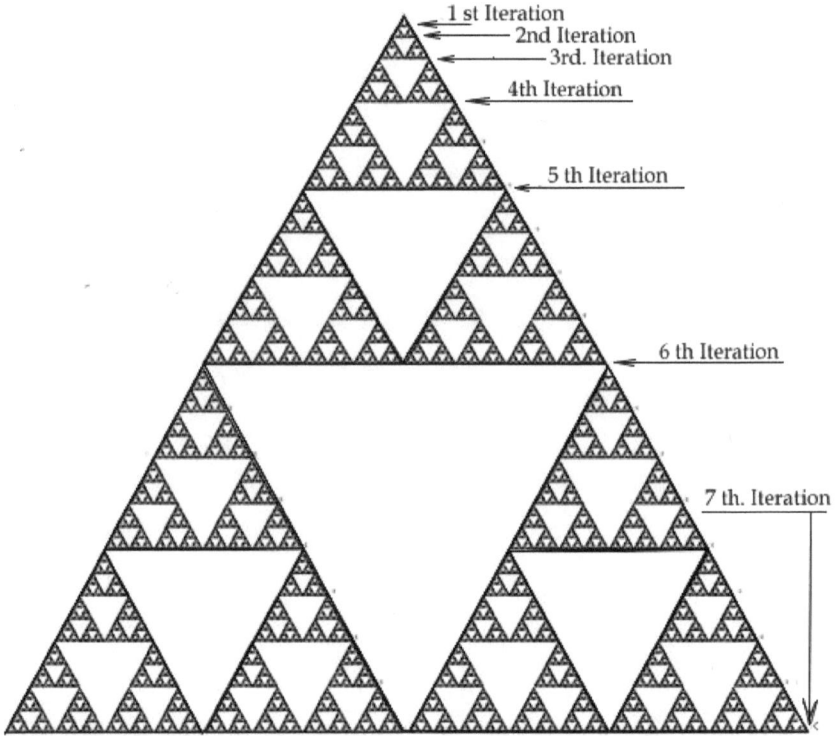

Fig. 1.1– The first seven iterations of the Sierpinski Triangle Fractal (STF)

skewed chiral universe according to Field Structure Theory. Three such loops interacting and properly knotting together produce a three-dimensional *fieldstructure* (Figures 1.2 & 1.3). The assumption is that all forms in the natural world are fieldstructures. To define a fieldstructure, a discussion on loops is needed. To do this, the Sierpinski Triangle Fractal is essential.

There are 2187 loops in seven iterations in the STF (Fig.1,1). This sequence is crucial to understanding the mass and energy hierarchy.

1st iteration has 3 loops
2nd iteration has 9 loops
3rd iteration has 27 loops
4th iteration has 81 loops

5th iteration has 243 loops
6th iteration has 729 loops
7th iteration has 2187 loops

Each of the seven iterations is a fieldstructure. Each field-structure is a particle. Particles are made of interacting loops While a particle is a closed structure that defines space, it should be thought of as a closed field of energy. As can be seen in the models, the loops are creating a nuclear polyhedron. It is this polyhedron that we call a particle, but as can be plainly seen, it is in actuality a form that arises from the interaction of loops. The loops are forming a field of action in which a polyhedron appears.

Fig. 1.2 – a computer generated fieldstructure by Josephy Clinton

Fig. 1.3 – an actual material model of a fieldstructure

Fig. 1.4 – trefoil - The Trefoil is not a true three-dimensional knot. It is a knot in three dimensional space, but not a knot defining 3-D space, because it does not create a polyhedron space.

Fieldstructures define a skewed three-dimensional space such as the tetrahedrons shown in Fig. 1.2 and 1.3 above. Both are *vortex* (vertex) views of a tetrahedron fieldstructure. Note: there is no glue or mechanical fasteners used in the Figure 1.3 working model. It holds its position simply by having the other two loops blocking each loop from being able to unbend and become stress-free circular loops.[2]

Three-dimensional interacting loops can produce all polyhedra, and any combination of polyhedra, however complex, extended and/or articulated. Any 3-D form can be built with loops. It is through interaction with itself or with other loops that loops quantify. If a form can define space, it can be made with loops. The implication is that all 3-D forms in the physical universe can be created by the interaction of loops. All that needs be done is to realize that electromagnetic energy is in the form of a loop, actually two entangled chiral loops. Now that we have a way of modeling all spatial form (with loops), the door is open to seeing if these loops can quantify themselves in the same manner nature quantifies energy, at each scale of complexity from particle to galaxy.

Loops can do two things.

Fig. 1.5 – a loop   Fig. 1.6 – Inside looping Fig. 1.7 – Outside looping

---

2   [Note: Previously "fieldstructures" were called "torsion structures".]

A loop can rotate about a distal center (a particular position, point, or place) and can do this in two ways (Fig. 1.5, 1.6 and 1.7) **via Inside looping and Outside looping.**

Figure 1.8 shares characteristics with an electromagnetic (EM) wave. There are similarities between an EM wave and a wave-set. The wave-set depicted in Fig. 1.8 is a fermion wave because the loops occupy distinct domains, while an EM wave is a boson wave wherein the waves are together in the same domain. Fermion waves cannot occupy the same space, but they can share the same axis, which is the point of contact between the two waves. In a boson wave-set, the two loops can occupy the same space.

Fig. 1.8 – **Wave-set,** Two loops rotated repeatedly about each other forms a shared axis that is the line of tangency where the two loops are meeting.

The two boson EM loops as well as the fermion wave-set loops in Fig. 1.8 share chiral relationships, among them is that one loop is rotating clockwise and the other counter-clockwise. Additionally, in the EM world, one is going forward in time, while the other is going backward in time. Real time goes forward and anti-time goes backward. The idea of the arrow of time is implicated, because we being real-matter only see the wave we experience as going forward. The anti-matter people see what is for us real-matter people going backward in time. This is fundamental to answering the question, "Why do waves wave?" EM is an entanglement of real-waves and anti-waves, which in turn set up the mechanism for matter and anti-matter to exist.

Natural electromagnetic (EM) loops have both clockwise (CW) and counter-clockwise (CCW) rotation. This pattern is found in smoke rings in air and air rings in water. In their deployed form (Fig. 1.8), which loop is CW and which is CCW cannot be determined. It is only by condensing one loop or the other does the issue of handedness arise. In energy (EM boson energy), the rotations are to be understood as occurring inside the same domain. They are inside each other. Fermion loops are outside of each other although entangled. For clarity, in Fig. 1.9 and 1.10, the loops are shown separated. A fermion loop has dimensions and can interact with itself as a single loop (Fig. 1.10), or two or more loops can interact (Fig. 1.9). The easy way to distinguish a fermion loop-set from a boson loop-set is to count the number of times the loops wrap around each other. All single loop-sets wrapped around itself have an odd number of nodes (odd number of twists), while all double loop-sets have an even number of nodes and are boson waves. Boson waves have spin 1 characteristics and Fermion waves have ½ spin characteristics.

Fig. 1.9 – Double loops rotating around each other have an even number of nodes. **Fermion wave**

Fig. 1.10 – Single loop rotates around itself and has an odd number of nodes. **Boson wave**

A loop has two basic attributes.
Looping
Twisting
These two processes of looping and twisting can perform.
deployment
condensement

In Field Structure Theory (FST) looping is associated with mass and twisting is associated with energy. In Field Structure Theory time/space and mass/energy are the same thing viewed from differing contexts and depend on each other for their existence. These two ideas are connected through the agency of field-structures. Mass is to space what time is to energy.

> **Field Structure Theory (FST) will show that the number of loops in a system is the mass number of the system and the number of 360-degree rotations of the loop (twists) around its axis is the energy number. Energy is dependent on the number of times the loop of action is twisted.**

This simplicity complies with the Occam's razor criteria for judging the worthiness of an idea by stating the simplest of competing theories is preferred to the more complex. In FST, that simplification means: all form and structure in the universe is reducible and understood as being the number of loops and the number of times those loops twist.

The kinetic energy of the loop is the number of times the loop rotates about its axis. The number of twists on the loop is the frequency of the loop. A rotation of the loops about the axis, twists the line of action. **A full 360-degree twist of the line of action accounts for one unit of energy, as does making a loop account for a unit of energy.** It is easy to see how twisting a line of action introduces energy into the line of action. Not so easy to see is that by making a loop, energy is introduced into a line of action. Making a loop, effectively twists the line of action a full 360-degree rotation. Experimentally, it can be shown that there is a way to make a loop with a piece of rope so that the loop will have a full 360-degree twist when the loop is formed. By bringing the two ends together, without bending the loop into a circle, forces the loop to rotate 360 degrees. _Note: Another way to see how space is connected in such a way as to cause twisting can be understood by coiling a hose into a circle._ To make a hose into circle and have the hose lay flat, without kinks, the hose has to be twisted 360 degrees each time the hose is coiled (looped).

This is a fundamental property of space that applies at all scales of structure (micro, macro, and cosmic scales). **Making a loop is, in essence, condensing space, and to condense space requires a twist to be introduced into the looped line of action.** Assuming (as we do) that space is the distance between two points, a property of space states that to close action into a loop, space has to be condensed/compacted. The consequence is that by condensing action, energy is created in the form of torsion (twisting). The number of condensing loops that are made and the number of times the line of action twists 360 degrees to make a loop quantifies energy in loops. In nature what is being looped is the Plenum. The Plenum is the substrate of reality.

The total energy of a loop is:
1. the number of loops
2. times the number of loops squared.

In Field Structure Theory (FST) the equation is:

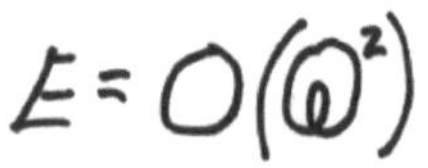

$$E = O\left(\omega^2\right)$$

Fig. 1.11 – FST's energy equation

This equation is the same as the equation used in physics: E = m(c^2).

The equation above uses a circle for the loop number, and the circle with a loop in it as the twist number. The twist number is squared. This suggests that energy is a relationship between distance and time and the number of times the line of action loops within that distance-time interval. The length of a loop, its circumference, is not affected by how big the loop is, because the loop is an instant no matter how big or small it is. What is at issue is the number of twists (nodes) in the loops of action. The energy is in the number of twists and is always going to be a relationship between the number of loops and the number of twists that loops have made. Planck's constant maintains that ratio.

The important thing about this way of looking at $E = mc^2$ is that mass and the speed of light squared are not separate things, but are two aspects of the same thing; twisting. **Loops inherently twist and twisting is quantitatively the same thing as looping.** If you know the frequency of the form, which is the number of times the form twists, you know the mass of the form, its loop number. Each loop number (mass number) has a specific twist number (frequency). Importantly, every fieldstructure has its own unique number of loops and twists. Each fieldstructure has its own unique frequency.

> **With these assumptions it is now possible to apply loop energy architecture to the mass and energy values of fundamental particles.**

The simplest form of interacting loops is shown in Fig. 1.2 & 1.3. These fieldstructures are composed of three loops, satisfying the quark model for fermion particles. The three loops gain energy by twisting the loop. As the loops add twist, they have parameters of the amount of twists the loops can have before the loop reaches a critical quantity of twisting, after which it jumps to a new loop configuration that can handle additional twisting.

Every time the loops "jump" to a more energetic loop structure, the loop increases in volume. The 2nd electron shell of an atom is a higher energy state than the first and has a larger circumference. While an electron jumps from one energy state to another, there is another process going on in the opposite inward direction that requires the energy to increase frequency and the volume to decrease such as when a *loopage* goes from being a deployed positron to becoming a condensed proton. An example is the positron loop. When it condenses, its spatial domain condenses 100,000 times. That increases the energy in units of electron mass from 3 units of mass to 1836 units of mass and the twist energy from 9 units of twist to 3,370,896 units. In other words, the proton loop is 3,370,896 times more powerful than the deployed positron loop, even while both are states of the same *loopage*.

Traditionally, physics has not had a quantum model that incorporates a charge field with the particle nucleon. Fieldstructures show that the action field external to the nucleus as the Bohr atomic model describes does not extend into the nucleus of an atom. FST reveals that a particle has a charge field associated with the nucleus of a particle in a similar way as does an atom have a charge field surrounding its nucleus and importantly, the field lines (the loops) extend, into and out of, the nuclear domain. The charge field of an electron is composed of neutrino loops. The charge field of an atom is the electron loops. More on this follows after taking a look at what actually is a real-matter fermion particle.

The standard by which real-matter fermion particles are known includes the following attributes. All these attributes are present in a fieldstructure.

1. Defines a 3-D domain of action.
2. Has a rest mass.
3. Has a minimum energy.
4. The nucleus of a particle is composed of three condensed left-handed clockwise quark loops and three right-handed

deployed loops. Of these two, one set of the three is visible real-matter and the other set of three is invisible anti-matter loopage. The Standard Model does not recognize that the anti-matter loops are fully invested and present though unseen in the particle's nucleus. The proof of their existence is in the fact that waves have frequency. The fact a wave pulsates is because the anti-wave is obscuring periodically the real-wave as the two waves rotate around each other.

5. In an atom, the deployed loops are behaving as electromagnetic particle waves (electrons).

6. In an atom, the condensed electromagnetic (EM) loops are quarks. Quarks are entangled and knotted loops that form in the nucleus. *The fact that the nucleus has polyhedron characteristics is not understood in the Standard Model of physics.*

7. The polyhedron geometry of a nucleon is made of condensed electromagnetic loops (quark loops).

8. The Standard Model postulates that a field particle called the *gluon* binds the nucleons together. In FST terms, gluons are a kind of nuclear valance particle, the way a valance electron in the atom can circuit with other atoms to form molecules.

9. The electron is a field particle of an atom and the nuclear particles are the atom's field object. The gluons, which are condensed neutrinos, are the field particles of the nucleus.

10. All particles have a field particle and a field object. The field particle is a deployed loop while the field object is a condensed loop.

11. Quarks deploy when the knot that holds them together is broken by the input of overwhelming energy, i.e., the *binding energy* is exceeded. Quarks that are deployed because of the binding energy being exceeded become gamma waves.

12. Fermions obey the loop quantum equation.

13. Condensed quarks have spin either clockwise (proton) or counter-clockwise (anti-proton). Real matter quark particles (protons, for instance) spin clockwise/left-handedly. Anti-matter nucleons spin counter-clockwise/right handedly.

14. The neutron has the deployed electron loop condensed on to a proton in the nucleus. This can only happen if another electron remains deployed.

15. A fermion can have a charge if either the clock-wise rotating loop or the counter-clockwise rotating loop of a wave-set is condensed, while leaving the other loop deployed.

16. Real-matter and Anti-matter are entangled. There is as much anti-matter as there is real-matter in the universe. Real-matter and anti-matter alternatively cloak each other producing a pulsing wave affect.

17. A fermion is entangled with a counter-part particle that is anti-matter, identical to real-matter except spin handedness and charge are reversed.

18. A fermion is composed of six loops, three of which are real-loop quarks and three are anti-loops quarks.

19. Real-matter cannot see (empirically contact) anti-matter when in the same quantum system. Even though entangled, they are invisible to each other.

20. Real and Anti-matter particles from separated quantum system annihilate if they meet.

21. A fermion pulsates and appears as a wave because the real-matter wave cloaks the anti-matter wave in a rhythmic pattern that produces frequency and pulsation as the two rotate around (by) each other (Fig. 1.40 & 1.43).

All particles have the basic form of a Structor as shown in Figures 1.2 and 1.3. It is possible to compute the mass and energy requirements of fundamental particles by counting the number of loops in the particle, which reveals the polyhedron of the form and the frequency, i.e., number of times the loops make a full twist in circuiting the form. The particle hierarchy of loops when diagrammed by the Sierpinski Triangle Fractal (STF) reveals mass and energy values in terms of loops and twists. Mass in particles increases by multiples of three. Energy in particles increases by the number of loops squared. Assuming a line in the STF is a quark loop, then the three lines of the triangle at the top of the STF are the three loops of a first-generation fermion. As the STF builds, the three quark loops become more energetic, which structurally mean the three loops are increasing the number of times they rotate by twisting. That number of twists reveal in the iterations of the STF, that particle mass increases in multiples of three. It is not that additional loops are being added, but rather the number of times the three quark loops continue to loop.

The mass value of the first iteration of the STF is 3 since there are 3 line/loops in the topmost triangle of the STF. The energy value is 9 (3 squared) since each loop rotates three times in completing the loop. This triangle is the smallest unit of fermion mass in the *Electron Sequence* of the STF. Since the electron is the smallest quantum unit of fermions, it has been used, and will be used for now, as the unit of measurement for all heavier particles (muon, meson, tau, proton, lambda, etc.).

However, a more accurate unit of measurement would be the neutrino's mass. Neutrinos are in a fractal hierarchy below the 1st generation. Hence in FST, the neutrino is the 1st generation making, the 2nd generation the fractal hierarchy that spans the electron to the lambda hyperon.

Meanwhile, using the Standard Models 1st generation classifications, all particles in the first generation of particles are multiples of the ground state electron having three loops and nine twists. The iterations of the STF defines the mass of the known stable and unstable (resonance) particles. In other words, a particle's mass will be in multiples of three quark loops. Particle mass builds in loop multiples of three. The illustrations in Fig. 1.12, 1.13, 1.14 show how looping increases in multiples of 3. Each figure is a single loop. When translated from the 2-D diagram shown below to a 3-D fieldstructure, the fieldstructures shown in Fig. 1.2 and 1.3 are formed.

Fig. 1.12 – Three sub-loopings on a single loop	**Fig. 1.13** – Nine sub-loopings on a single loop. Note the over under sequence is incorrect.	**Fig. 1.14** – Twenty-seven sub-loopings on a single loop

Fig. 1.12, 1.13, and 1.14 shows what is meant by multiple looping in multiples of three. The STF shows the loops as being separated in a progression of multiples of three, whereas in Fig. 1.15, the loops build as a continuous progression of integers (1, 2, 3, 4, etc.). In 3-D it is not possible to build structure as a continuous progression; the <u>Multiple 3 Rule</u> applies in 3-D. It takes the STF to show why masses do not build continuously, but rather have to occur in multiples of three (3, 9, 27, etc.). If fermion mass were continuous, the sequence would be a simple arithmetic progress. The fact that fermion loops do not progress arithmetically (1, 2, 3, 4, etc.) is because when energy interacts to form a fieldstructure, the progression has to be in multiples of three (the 3, 9, 27 Rule) in order to have structure.

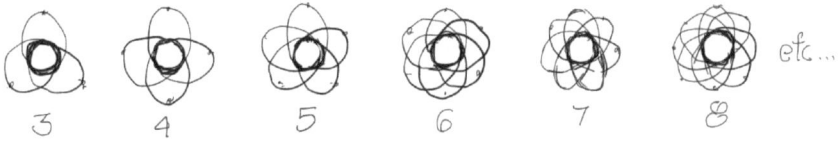

Fig. **1.15** – An arithmetic progress of loops can be found in boson loops (EM waves), but not in fermion loop structures.

The electron having three (quark) loops is the unit of measurement within the Standard Model's first-generation. Heavier particles will be multiples of three of the electrons and will be divisible (decay) by three.

Therefore:

The first iteration of the STF will be 3 loops.

The second iteration of the STF will be 3 times 3 for a total of 9 loops.

The third iteration of the STF will have 3 times 9 for a total of 27 loops.

The fourth iteration of the STF will have 3 times 27 for a total of 81 loops.

The fifth iteration of the STF will have 3 times 81 for a total of 243 loops.

The sixth iteration of the STF will have 3 times 243 for a total of 729 loops.

The seventh iteration of the STF will have 3 times 729 for a total of 2187 loops.

The seventh iteration contains all the loop mass needed to account for the first-generation of particles AND have the same energy as found in hydrogen.

2 - Summary

FST proposes that a particle is made from entangled chiral loops that in nature are right and left-handed electromagnetic loop/waves. The entangled loops form a *loop-set,* also called a *wave-set.* **The electromagnetic wave-set is composed of an *electron-wave/loop* and a *positron-wave/loop.*** When three such chiral loop-sets entangle at the correct frequency, one side of the entwined loops condenses to form a nucleon. The other side of the wave-set remains deployed and becomes the charge field. The condensed nucleon and the deployed field particle are inseparable. The condensed wave structure forms a polyhedron field object and the deployed wave becomes the charge field. Each particle has a unique wave frequency associated with it that is determined by the polyhedral structure of the nucleon.

The electron has internal structure. The electron's nucleus has 99.94% of the particle's energy and the electron's charge field has .06% of the energy. The electron's charge field has three *tau-neutrino* loops. The positron has for its field particle the *anti-tau-neutrino.* The wave and the particle are inseparable just as an iceberg and the ocean are inseparable and made of exactly the same stuff, the only difference being their energies. As an iceberg goes where the ocean currents takes it, so does the electron and positron go where their respective electromagnetic waves take them. It is the nucleus that makes that determination. This is the structural origin of de Broglie's pilot wave concept.

All distinctions such as quark, particle, wave, gluon, photon, etc., are the varying states of loop condensement. All things have loop structure. At the fundamental scale, an action loop is electromagnetic (EM). Everything in nature can be reduced to EM energy, and even energy can be subsumed by the Plenum. The Plenum is the source of energy. The effect EM energy has on the Plenum shows up when EM energy knots into a particle having mass. The stress induced in the Plenum by the formation of mass is the force of gravity. When action is condensed, the Plenum contracts. When the Plenum condenses, space compacts.

Space contracts when energy is condensed.
Space expands when energy is deployed.

The idea that space does anything, such as condense or deploy, is generally not in the purview of physics. The idea is fundamental to FST.

The energy to create mass is drawn from the Plenum. To understand gravity, the effect of producing particles from EME has to be considered by looking at how EME is made from Plenum. Making mass puts stress in the Plenum. That stress is gravity. The Plenum does not want to be condensed.

Starting with the loop, all forms in nature can be derived depending on how a loop interacts with itself or other loops. The substratum of form is the loop.

FST makes the following assumptions:
1. Fermions are composed of three loops.
2. Each loop is composed of entangled clockwise (CW) and counter-clockwise (CCW) loops.
3. When those loops are maximally condensed, quarks form.
4. A line of action rotates 360 degrees to make a loop and the line rotates around its axis 360 degrees is a count of looping and twisting.
5. Total energy of a particle is a count of the number of times the particle rotates both in terms of loops and twists.
6. A mass form appears at each loop hierarchy in the STF.
7. Electrons can take on energy, iteration by iteration.
8. Each shell of an atom has a minimum/maximum energy distribution.

9. FST considers the electron as a fieldstructure unit of inter-acting loops and twists, rather than a particle having no discernible internal structure. Particles are fields, not a point particle.

10. FST considers the whole of reality to be action loop fields of one degree of condensement or another.

11. The view that particles are separate from the Plenum[3] and background independent is not supported by FST.

12. FST sees nuclear fermion particles as three interacting EME waves (3 quarks loop/waves) and the electron and positron as three interacting *spark* loop/waves.

13. Spark loop/waves can condense to the nucleus and become quarks, but quarks cannot deploy to become *spark* loop/waves. Quarks cannot become sparks because once the quark's knots are broken, they are released as gamma radiation which is too powerful to become a spark.

14. Since the process of particle creation is not reversible, nature creates mass by first building a particle that has little, possibly no, kinetic energy, i.e., a static particle. This is the process whereby the lambda is formed by EM interacting at super-cold temperatures. By decay-ing cold, the particles that are produced by the decay have little kinetic energy. This will allow EM interaction between the particles to bind them together. The proton can capture the electron and create hydrogen. EM energy becomes particles and particles become atoms through cold fusion.

15. The structural form three loop/waves (quark) have when they interact is a fieldstructure.

16. The number of loops and twists determines the particle.

---

3   Plenum is a cousin of ether (or aether), related but different.

17. Each iteration of the STF above the first iteration is an unstable energy state whereby the loops have a momentary semblance of structure before decaying.

18. These unstable states, when integrated into an atom, have energies the same as found in the shell of an atom. Nature has provided a way to stabilize unbound and unstable particles by bounding the energy into a fieldstructure.

19. For instance, the muon is unstable, but that same amount of energy in the 5th electron shell of the atom is stable. The difference between a muon and the electron in the 5th shell is that the former is an unbounded free ranging muon particle while the same particle bound up in atomic structure is an electron in the 5th shell.

20. Atoms can have up to seven shells. Each of the seven shells has the energy found in the seven iterations of the STF. Electrons bound to an atom in shells have the same energy as are found in the seven particle iterations of the STF.

21. The seven iterations of the Electron Sequence define first-generation particles in the Standard Model.

22. The energies of the seven shells of an atom and the seven iterations of electron Sequence are the same.

23. Energy that is bound up in the seven shells of an atom is stable. The same energies when outside the atom in free space are the unstable resonance particles.

The problem with seeing particles as separate entities is that the three- loop structure of fermion particles is not apparent. As loops condense, they compact. As they compact their frequency and hence energy increases while their spatial volume decreases. The loop form when deployed has the form of the C-FS (Clinton-fieldstructure) (Fig. 1.16). When the loops condense, they have the form of the WB-FS (Wysocki/Briddell-Fieldstructure). Both forms have the same number of loops but define unique spatial forms.

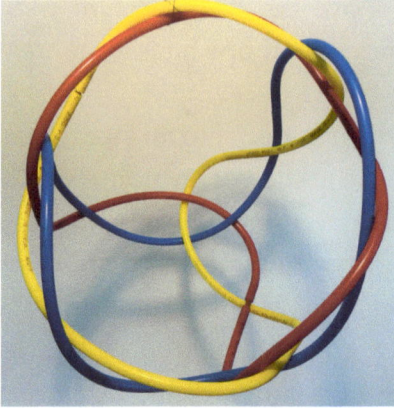

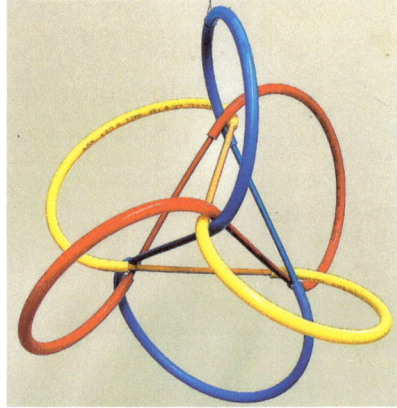

Fig. 1.16 –Clinton-Fieldstructure	Fig. 1.17 – Wysocki/Briddell-Fieldstructure
Three loops knotting deploy to form a spherical space when deployed. Loops are all on the surface of a sphere. Vortices, where loops interact are on the sphere's surface.	Three loops knotting condense to form a nucleated sphere. Loops circuit the interior of the forms volume. Vortices are in the interior of the sphere.

Surprisingly, though for a good reason, the proton and neutron are not iterations of the STF. Six of the seven iterations of the STF are *resonate particles* because they exist only momentarily before disassembling and becoming either a simple electron(s), or neutrino(s), or radiant energy, or all of these things. These resonant states have names. The 5th iteration resonate particle is called a muon. The 6th iteration is a resonant particle called the tau. The 7th iteration particle is called the lambda hyperon (*hyperon* means it is a particle that is heavier than a proton).

The 1st generation of particles ends with the lambda, because all Standard Model first generation particles can be made from the energy in the lambda. The lambda can produce in its decay scheme a proton, an electron, and a neutrino. If lambda decay occurs in a hot environment, such as in a collider, the decay products will be highly energetic and unable to associate with each other. No atomic structure can form. If the decay occurs in a cold environment, the decay will form hydrogen. In cold decay, all the needed ingredients of an atom are present and cold enough for the magnetic attractive forces between the positive proton and the negative electron to operate and bind the two particles into an atom!

There are hyperon particles heavier than the proton that are found in the 2nd and 3rd generation of Standard Model particles. This paper will not explore these hyperon particles since they are all unstable and decay to first- generation particles and/or electromagnetic energy. That task will be left to a future paper.

Given these assumptions, the following can be said of the loop hierarchy:

1. All energy is in the form of interacting loops.
2. Fields are condensed loops.
3. The degree of *condensement* determines the wave form of its iteration state and particle associated with that particular energy state.
4. Interacting loops have degrees of incremental non-linear condensement.
5. Quarks are maximally condensed electromagnetic loops squeezed into the nuclear dimensions and form nucleons.
6. Quark baryons loops cannot deploy to the charge field of an atom and become the spark loops of leptons, because quark loops are too energetic to form loops of a lower energy particle such as spark loops.
7. However, a spark loop can condense to become a quark loop.
8. Lepton loops originate from the decay of lambda energy particles.
9. Loops can loop many times as shown in Fig. 12, 13, and 14.
10. A deployed quark wave is a radiant gamma ray.
11. A deployed spark wave (lepton) is a radiant X-ray energy.

12. The number of times a loop loops in a period of time determines the mass of a particle, just as the number of times a wave waves determines its frequency (energy).
13. The unique thing about field structure loops is that they quantify energy and do so hierarchically.
14. If loops quantified continuously there would be particles for each number of loops (1, 2, 3, 4 etc.). In FST, a continuous progress is one-dimensional, not found in a three-dimensional world. A continuum in 3-D increases in increments of three.
15. Natural 3-D loops quantify hierarchically. They quantify in multiples of three (3, 9, 27, 81, 243, 729, 2187, etc.)

Loops deploy or condense and the results explain the particle hierarchy. Waves form particles depending on how much the Plenum has been condensed. Particles form when the condensation of the wave reaches a frequency sufficiently strong for the EM forces of attraction and repulsion enable the loops to knot. The knot forms when the tipping point is reached. When three such waves form a particle, the forms become a mass structure having a proscribed limited domain.

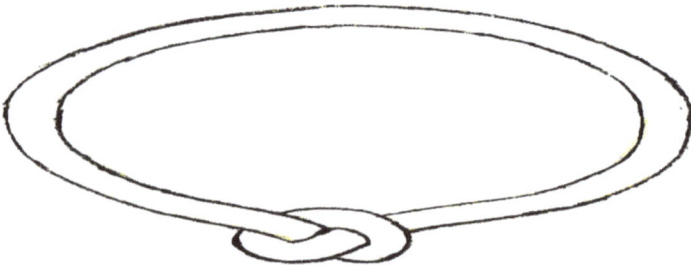

Fig. 1.18 – Knot on a string: FST considers the knot to be the *field object* and the *action field* is the string. Every field object has an action field associated with it, just as a knot is also a string. The field object is where a loop interacts either with itself, or with other loops.

3 - Loop Families

Spark loops condensed become quarks.
Quark loops deploy to become gamma rays.
Spark loops deploy and become X-rays.
Lark loops condense to become Spark fields.
Lark loops deploy to become the mid-range EM waves.
Ark loops condense to become Lark fields.
Ark loops deploy to become light and radio waves.
All loop families can deploy to the Plenum.

<u>Note:</u> *The Plenum, in FST, is called the "quantum loop field", the same term as used in Loop Quantum Gravity physics.*

In this vision of how nature is structured, all forms of nature are derived from loops. Thus, all quantitative forms are understood in terms of the number of loops and the number of times the loops rotate (twist).

The following applies:

1. Each of the lines in the STF represents a loop.
2. A loop is electromagnetic energy (EME) and the energy of the loop is determined by the number of times the loops rotates around its axis.
3. There are multiple degrees of loop condensation, the most intensely condensed loop form is the quark.
4. Deployed quark loops are not seen, because once the nuclear knot is broken, the loops become radiant gamma loops. To condense back into quarks three gamma wave/loops have to interact at the right frequency.
5. Three EME loops condense and automatically knot electromagnetically when they interact at nuclear dimensions to become quarks.

6. The condensing mechanism is in the electromagnetic properties of attraction and repulsion. Fieldstructures are how nature achieves equilibrium between centrifugal and centripetal forces. When chiral EME loops interact, the process of condensement automatically forms a three-dimensional knot called a fieldstructure.

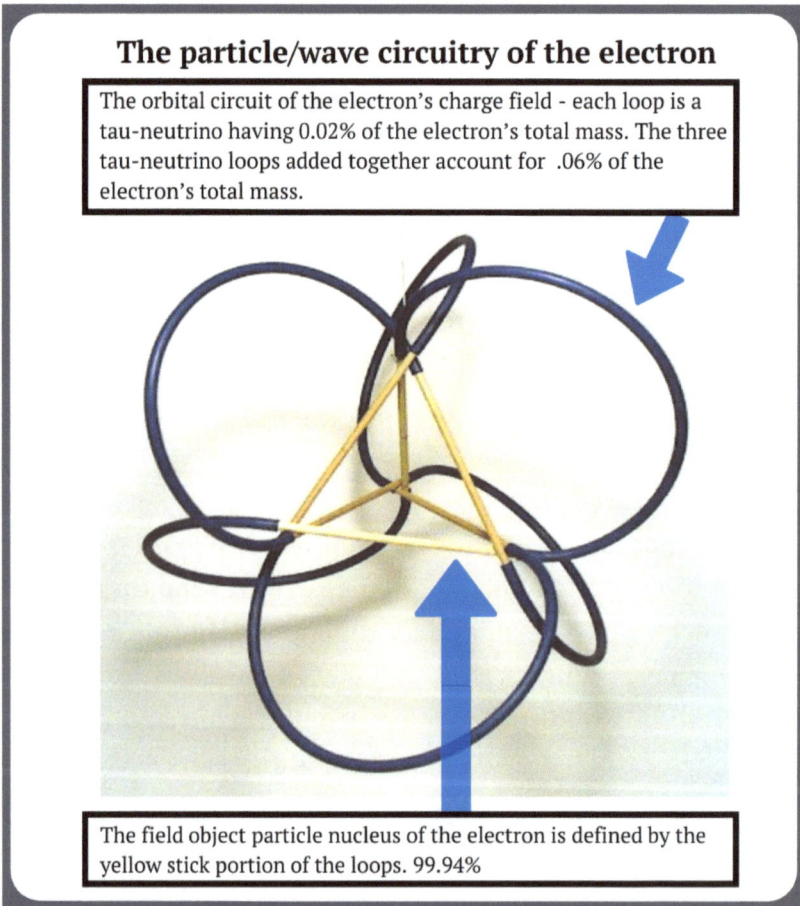

The particle/wave circuitry of the electron

The orbital circuit of the electron's charge field - each loop is a tau-neutrino having 0.02% of the electron's total mass. The three tau-neutrino loops added together account for .06% of the electron's total mass.

The field object particle nucleus of the electron is defined by the yellow stick portion of the loops. 99.94%

Fig. 1.19 – Structor Fieldstructure: three loops interacting topologically forming a three-dimensional knot. This is the architecture of particles. The simplest particle forms a tetrahedron nucleus. The figure shown is called a Tet-Structor.

4 - Analyzing the Sierpinski Triangle Fractal

Traditionally, the electron has been the base unit of the STF's **Electron Sequence**. The electron sequence has seven iterations. These seven iterations include all fermions of what the Standard Model calls the first generation of particles, such as hadrons, mesons, muons, electrons and etc. FST calls the electron sequence as the second-generation, because there is a sequence with masses below the electron-sequence that has the same sequence of particles but at lower energies. That lower sequence is the **Neutrino Sequence**. The base unit of measurement for the **Neutrino Sequence** is the electron-neutrino. It is a sequence of loops identical to the particle loops in the electron sequence except the particles have a smaller mass.

The electron is composed of three spark loops, just as the proton is composed of three quark loops; the difference being the loops have a different frequency. The proton nucleon is thought to have three quark loops, but the electron is thought of in the Standard Model not to have an internal loop structure. FST postulates that the electron has the same three loop structure. The difference is the electron loops live in a larger volume of space. FST claims the three loops in the electron are tau-neutrino loops. As you will see in this book, the mass numbers bear this out. The neutrinos connection to electron structure is not known structurally due to its tiny mass and the fact that once liberated from the electron, it is nearly impossible to detect. The same structural configuration is carried through to all scales of structure.

The only way presently to verify this is by uperating the theory to where empirical measurements can be made. Experimentally, physics has determined these numbers, but until this thesis, it has had no way of relating loop families to each other, due entirely to the fact the connection between particles and loops has not been established.

By the seventh iterations of the STF, there are 2187 loops (Fig. 1). As shown previously, the loop sequence is 3, 9, 27, 81, 243, 729, 2187. In these numbers, and combinations of these numbers, all particle masses found in the first-generation of particles can be generated. All iterations of the Electron Sequence are resonance particles that are highly unstable, except for the electron. The proton is not an iteration of the STF. The neutron, while stable as a nucleon, decays if removed from the nucleus. By sharing an electron with a proton, the neutron remains stable.

The **Neutrino Sequence** has three combinatory forms:
1. electron-neutrino
2. muon-neutrino
3. tau-neutrino

It takes three tau-neutrino loops to make an electron. The three loops of the electron's structure in Fig. 1.19 are tau-neutrinos. By interacting, tau-neutrinos condense the positron wave side of each wave-set to form the nucleus. On condensing, the positron wave takes all the energy of the wave-set with it except for one unit of twist left in the three deployed wave/loops.

The **Electron Sequence** has three combinatory forms:
1. electron.....(first-generation field particle)
2. muon ....(second-generation field particle)
3. tau...........(third-generation field particle)

There are seven iterations in the Neutrino Sequence and seven in the Electron Sequences. Both have three combinatory forms. Combinatory forms occur when three energy iterations of the STF combine.

Three electron-neutrinos make a muon-neutrino.
Three muon-neutrino make a tau-neutrino.
Three tau-neutrinos make an electron.

An electron is a combination of three tau-neutrinos loop structures.

A muon is a combination of three electrons loop structures.

A tau is a combination of three muons loop structures.

The Neutrino Sequence is as yet unrecognized by physics. In fact, the whole concept of "sequences" is new to physics. By looking at particles as loops, the mass of particles becomes clear. The three-loop hierarchy of structure may exist all the way down to a theoretical Planck particle. Physics to date has seen experimentally no mass structure below the neutrino. The Large Hadron Collider in Europe is looking for sub-neutrino fermion particles; the Higgs particle that they assume is a boson. FST assumes there are fermions smaller in mass than the electron-neutrino, probably several billionth of the electron's mass. For now, such particles lie outside of experiment verification.

Anti-neutrinos have the same structure as real-neutrinos, but the loops spin in the opposite direction. This is known by standard physics. Neutrinos are the lark loop family of form. Lark loops condense to form denser spark loops. Spark loops are leptons. Leptons are electrons. Condensing spark loops produce quark loops. The Standard Model of physics considers the neutrinos to be a first-generation particle. In FST, neutrinos are first generation particles and electrons through Lambda are 2nd generation particles. This is done not to be contrary, but is done to show how FST interprets all particles as loops arranged in fractal hierarchies.

Not done, but interesting to do, would be to build a STF that has 21 iterations which would account for all know 1st, 2nd, and 3rd generation particles. Beginning with the electron-neutrino there would be 28 iterations between the electron-neutrino and the heaviest of the fourth-generation of particles. That would be 3^{28} or 68,630,377,364,883 electron-neutrinos in the heaviest of the fourth-generation particle.

The Neutrino Sequence in FST is the first-generation of particle fields. The total energy/mass of all seven iterations of the Neutrino Sequence constitute the first particle field of the Electron Sequence, which is the second-generation of particles. The Electron Sequence has two stable particles, the electron and proton, as well as the resonance particles such as meson, pions, koans, tau, etc. The neutron is a combination of proton, electron and neutrino particle fields.

Given our limitation of seeing and working with what the Standard Model calls "the First-generation of Particles" that begin with the electron and end with the Lambeda, and given the fact there are known particles beyond the lambda's mass and particles below the electron's mass, we have s similar situation in EME where we "see" only the visible light spectrum while we know there exists light at higher frequencies and lower frequencies than what we can "see". The same phenomenon exists with particles.

Beyond FST's second-generation, which is the Electron Sequence, there is the third and fourth-generations of particles, but not a fifth-generation. While the second-generation Electron Sequence has the electron with 3 loops as its base unit, the third-generation has the muon with its 243 loops as its base unit and the fourth-generation has the tau with 726 loops as its base unit. The third and fourth-generation particles can be made in the laboratory, they do not exist outside of high-energy collision environments, except when cosmic particles crash into second-generation particles (electron, protons and neutrons). It may be possible for 3rd and 4th generation particles to exist in stars.

The fractional values of masses in the Electron Sequence when measured in neutrino masses instead of electron masses gives a more precise number for particle mass. The degree of certainty is increased 2.2 million times greater than when using the electron to determine mass. By using the neutrino as the unit of measurement, the fractional values established in physics when measured using the electron's mass (or energy) as base one is explained and justified. For example, the proton in electron masses is 1836.15267245. The 0.15267245 fraction in the mass of the proton occurs when measured in electron units. The question is, "Why is there a fraction?"

FST's answer to that question is that physics has been using the electron to measure mass and did this by assuming the electron has no internal structure, and no field particle (the neutrino) associated with it. The fractional value of the proton, results from not considering the extra mass of the electron-neutrino.

By using the energy of the neutrino as base one, instead of the electron, a much more accurate number can be used. Looking at the proton, the 1836 number is the mass contribution of the electron's nucleus and the 0.15267245 is the tau-neutrino's mass contribution outside the nucleus in the charge field of the electron. Assuming there are 1836 electron mass units in a proton, then there are 1836 units of tau-neutrino units as well. The question becomes, "What is the mass a single electron-neutrino and what mass would 1836 electron-neutrinos add to the mass of the proton?" Since the electron-neutrino field particle is structurally inseparable from the electron, what amount of mass does the electron-neutrino add to the mass of the electron so that mass of the proton becomes an even number with no fractional values?

5 - Using the Neutrino as the Unit of Measurement

This has not been done in physics, because the mass of neutrinos has not been sufficiently determined. Using Tometry, this is now possible.

As previously mentioned, the problem in physics is to understand why using as the measuring unit the smallest mass in the Standard Model's first -generation of particles, i.e., the electron, the heavier particles will have a fractional value? Why is that? One would reasonably assume that if the electron is the smallest particle, all other particles would be a whole number when measured in electron masses. The electron is not a primary particle or the wrong mass is being used for the electron. I favor the view that both are faulty. The electron mass has not been properly measured, nor is the electron a primary particle in the hierarchy of particles.

The problem is the field particle associated with the electron has not been included in the mass of an electron and has not been seen as the fundamental brick in the wall. The field particle is the electron-neutrino and it has mass. In this paper, it will be shown that when the electron-neutrino field particle is added to the mass of the electron, all the other particles will have a whole number value with no fractions. This, of course, assumes the neutrino does not have internal structure. However, in a fractal universe, even the neutrino has internal structure and would be a composite particle as are all other particles. While the electron-neutrino is tiny, whatever makes up the neutrino would have a mass of 0.0000000000915 which is 2187 times smaller than the neutrino. Counting mass with the electron-neutrino will give us enough precision for the foreseeable future.

As has been mentioned, FST considers the first-generation of particles to be the **Neutrino Sequence (NS).**

Recognized by physics, there are three neutrino particles, each with their own energy value.

1. Electron-neutrino
2. Muon-neutrino
3. Tau-neutrino

These three neutrinos are associated with the heavier electron, muon, and tau particles.

FST will measure – for now – with what physics knows to be the smallest particle having a mass-locality, the electron-neutrino. The problem with using the neutrino presently is that its mass has not been emphatically determined. The three neutrino particles all have different masses (energies) and it seems to physics these particles morph into each other for unknown reasons. Complicating the effort to measure the mass of neutrinos is the fact that they are speed of light particles and yet have a mass, albeit tiny. According to special relativity, a particle going the speed of light would have an infinite mass. Either the neutrino is not a speed-of-light particle, or if the neutrino is not a particle. FST will attempt to bring some order to this elusive neutrino particle and its role in particle physics by structural means. Consider the following table:

> **FST Hierarchy of Particle Sequences** – structurally these forms are loops within loops, each loopage forming a bigger fractal sequence.
>
> Neutrino sequence – FST's First-Generation
> > Electron-neutrino = single neutrino loop
> > Muon-neutrino = two neutrino loops
> > Tau-neutrino = three neutrino loops
>
> Electron sequence – FST's Second-Generation
> > Electron particle = three Tau-neutrinos loops
> > Muon particle = three electrons loops
> > Tau particle = three muon loops
>
> Lambda sequence – FST's Third-Generation
> > Lambda particle = three Tau particles. Hyperon particles in this sequence have-not been organized as yet beyond the lambda hyperon.

The tau+ particle in the Neutrino Sequence has 2187 loops. (Note: *The Tau+ designation means three tau-neutrinos*). *A single tau has 729 loops and is in Neutrino Sequence mass units. Three tau loops have 2187 loops.*) The electron is composed of three tau-neutrons. The lambda particle field has 2187 electron loops. There are seven iteration in a sequence and each sequence is a platform of structure.

The Electron Sequence begins when three tau-neutrinos from the Neutrino Sequence (729 times 3) interacting to produce the electron. Complicating the task of measurement is the fact physics has not counted the contribution of the neutrino into the mass of the electron. It measures only the particle's field object, which is the electron, and not the field particle associated with the field object, which in the electron are the three tau-neutrino loops. Known to physics, the hydrogen atom has 99.94% of its mass in the nuclear field object, the proton, and 0.06% of the its mass in the electron, the field object. The same relationship of field object to charge field found in atomic structure is found in particle structure.

The Sequences of Structure in FST beginning with the neutrino are the:

> Neutrino sequence = first-generation particle/fields
> Electron sequence = second-generation particle/fields
> Muon sequence = third-generation particle/fields
> Tau sequence = fourth-generation particle/fields

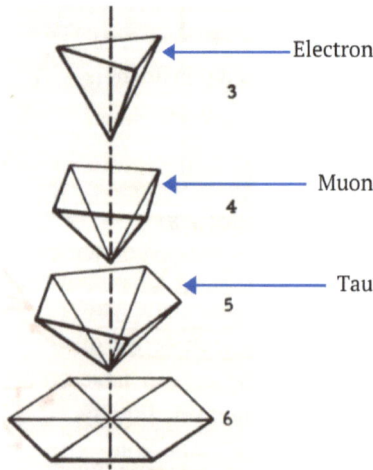

Fig. 1.20 – Closure constraints. Images 3, 4, 5 if repeated enclose space as do particles with these geometries. 6 does not enclose space – hence can't form a particle. Illustration from "Synergetic" by Buckminster Fuller.

Beyond the tau sequence (the fourth-generation) energy cannot form into a bounded 3-D particle field, because six triangles make a flat plane and the form will not close. It takes three loops to make a tetrahedron, four loops to make the octahedron and hexahedron (muon) and five loops (tau) to make the icosahedron and dodecahedron. There is no symmetrical stable polyhedron with six or more edges. This is why there are only five Platonic solids.

The images in Fig. 1.20 define the geometry of a particle generation.

1. The 1st generation of particles have three loops and the unit particle is the electron having a triangular structure.

2. The 2nd generation of particles have four loops and the unit particle of that generation is the muon and would have either octahedron or hexahedron structure.

3. The 3rd generation of particles has five loops and the unit particle of that generation is the tau and have either icosa-hedron or dodecahedron structure.

There is no 4th generation of particles, because such particles would not enclose space. Particles must enclose space to be a particle. The 4th generation exists, but not as particles. It would be a plane that has no closure and that is a definition of the Plenum. The 5th generation of energy, and any generation higher than the 5th , would be a wave although I have not built a 5th generation particle and am only speculating there would be no particle form in a higher generation; maybe at some point it would close. Undoubtedly, there would be surprises in what happens to the form as the number of triangles are added beyond the 5th generation. The form may tessellate. Speculating, it may lead to the reason why matter condenses when the energies are sufficiently high.

Curiously, and in keeping with the loop nature of reality, the fifth-generation does exist, but has no frequency; no frequency means no energy. However, the sixth-generation and all generations beyond the sixth-generation have frequency, because to add an additional triangle to a flat plane will cause it to undulate (wave).

Platforms of structure (different from sequence of structure).
1. wave
2. particle
3. atom
4. molecule
5. cell
6. organism
7. plant
8. animal
9. humans

Each platform is a fractal *uperation* of an adjoining smaller fractal. Fractals build when the total energy of one fractal sequence becomes the minimum energy event in the adjoining larger fractal sequence. As mentioned, physics measures the field object mass of particles that contributes 99.94% of the total mass, and does not include the mass in the deployed charge field that contributes 0.06% of the total mass. Both measurements are needed to have a complete mass accounting of a particle. Using only the field object mass will result in fractional mass values of any particle measured with the value of a particle's field object mass. Hence, when the mass of the proton is calculated using only the field object electron mass, 0.06% of the energy/mass that goes into making a more massive particle is missing.

Measuring with the electron's field object having 99.94% of the mass as base one, the proton will have 1836.15267245 electron masses according to the Standard Model of physics. The problem here is the electron being measured as base one, does not include the 0.06% mass of the electron-neutrino (E-N) field particle. If the E-N field particle is included with the field object particle then its additional mass will be factored into the calculation of proton mass. By including both masses, the proton particle will be a whole number. Since there are 1836 electron masses in the proton (discounting the fraction for the moment), by adding in the mass contributed by the E-N to each electron mass, the mass number for the proton will be a whole number of 1836 and not 1836.15267245. Not having the fraction gives a clearer picture of what is happening structurally.

According to FST, the fractional value of .15267245 in the proton mass (1836.15267245) is attributed to the added mass provided by the neutrinos (E-N). In the FST accounting, there are 1836 electron masses in a proton, thus if each electron has three tau-neutrinos (E-N) associated with it, then .15267245 divided by 1836 = 0.000083154929194, which would be the mass of the tau-neutrinos. **That means the new mass number for the electron is not one, but 1.000083154929194.** Thus 1.000083154929194 multiplied 1836 times equals 1836.152672451331, which is the known mass of the proton. <https://EN.wikipedia.org/wiki/Proton>. A single tau-neutrino would be 0.000083154929194 divided by 3 (the three neutrinos loops that make up the electron's field) gives **a single tau-neutrino mass of 0.000027718309765.** According to FST, this is the mass of a single tau-neutrino. A single electron-neutrino would then have a mass measured in electron masses of **0.000009239436+ (0.000027718309765 divided by 3).**

6 - The Proton as a Loop Structure

The Tometry analytic combined with Field Structure Theory reveals from whence particles get their respective mass and energy values. It is more than interesting that the 1836 loop number of the proton does not appear as an iteration in the STF. It would seem then that using loop counts to determine mass values does not work. However, nature has worked out a clever way to develop the mass hierarchy and build atoms from particles. It is not the physics of the Standard Model. It may seem reasonable to assume the electron and proton would frame the Standard Model's first-generation of particles (second-generation in FST). However, the second FST generation ends with the more massive **lambda hyperon** having 2187 electron loops, which is heavier than the 1836 loops of the proton. When the lambda particle's energy is measured, the lambda has the energy of 4,782,969 (2187 squared) using the formula: energy equals loops (2187) time loops squared (2187^2).

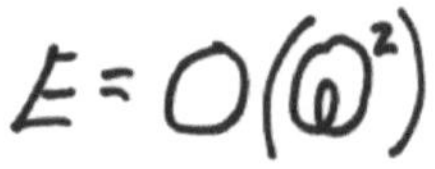

$$E = O(\omega^2)$$

Fig. 1.21

Nature is energy. Each iteration of the STF is where the boson energy can be localized to produce a particle, all of which are unstable and decay, except for the electron, remembering that the proton and neutron are not iterations of the STF. The muon, tau, koan, pion, and lambda are examples of unstable energy iterations of the FST"s second-generation (which is the first- generation in the Standard Model of physics). It is only when nature creates the lambda hyperon that the proton can be formed.

Nature seems to be saying it does this in order to create the other necessary components of matter, the electron and the proton, without the need of a thermonuclear reaction!

> **By creating the electron and proton as the decay products of a more massive particle, the lambda, thermonuclear fusion is not necessary. The creation of the electron and proton is a cold decay process, not a hot fusion process.**

In the main, fusion is not how nature creates particles as is presently assumed. While fusion can happen and has a place in the natural order, nature does not use it to make particles. Particles are not the result of thermonuclear fusion. Fusion in physics is a process that can be commenced once particles are formed. Cold fusion is what makes particles and atoms. While hot fusion occurs at the atomic level, it is not the primary way nature makes particles. Of course, this view is diametrically opposed by the astrophysicists

In FST, particles have to be created in a cold environment and for a good reason. In the Standard Model of physics, extreme pressure of fusion is necessary to fuse hydrogen into helium (a fusion reaction). This idea works, but it does not account for particle production nor for hydrogen production. The process only applies to atoms, not to particles. While heat and pressure can fuse hydrogen into helium. Heat and pressure cannot bring an electron to bind with a proton. Heat is antithetic to particle formation. Electrons and protons have to be cold for them to bind; not only to bind together, but to form from EM energy in the first place. Particle production must occur in an extremely cold environment.

In cold fusion, the heat is taken out of the equation so that EM energy can do its thing (so to speak). EM energy has no intrinsic heat. It is only when an EM wave impacts a mass that heat arises. Thus, the energy needed to produce the lambda particle can be obtained without heat. The lambda particle will form in a cold environment from gamma frequency radiation. However, the lambda is energetically unstable and as soon as the particle is formed due to EM gamma interaction, it decays into a proton, electron, neutrino and a meson. Not seen, the anti-lambda also decays into anti-particles at the same time the real-lambda forms real-matter particles. The proton and electron being cold, attract each other electromagnetically and form hydrogen. This is what is going on in the interior of the stars and planets. Yes, planets and starts have cores void of matter at extremely cold temperatures where cold particle production can occur. Structurally speaking, all LMO (Large Massive Object) stars and planets have cores void of mass into which mass cannot enter. Why LMOs have cores that are void of matter is another discussion I've dealt with in a separate paper.

Particle formation has to be a cold process, so that EM energy can interact without the presence of mass. Mass carries heat so it must be absent. Once hydrogen is available, and put into a hot environment, larger atoms can form through fusion. That process occurs in the super-heated corona of a star 100,000 miles from the surface of a star. The sun's matter body, the part we "see", in the grand scheme of things, is relatively cold. The heat of the sun hot enough to cause fusion, is in the corona surrounding the sun's mass.

What is needed to create atoms is an environment where high-energy gamma rays can interact three-dimensionally in a cold environment in a space void of matter, to produce a particle. Crucial to understanding this process, is the fact that high-energy gamma rays are not hot in their radiant form. It is only when these rays impact a mass, that the mass experiences heat. The term "cold energy" sounds like an oxymoron, but it isn't. Heat is a result of EM waves impinging on matter. Until the wave impinges on matter, there is no heat. EM gamma energy can interact without heat and condense into matter.

Three EM gamma energy loop/waves each having 729 loops form momentarily a lambda particle having 2187 loops. The lambda would be cold since it formed from EM energy which in itself has no heat; there is nothing in EM energy to heat. The sun's corona becomes hot only when mass is ejected from the star's mass and tossed into the corona. The sequence would be: (1) cold at the sun's core (at or near zero), (2) hot on the surface of the sun (4,000 to 6,000 degrees), (3) super-hot by 100 million degrees in the corona.

In open space, EM radiation does not interact with itself or other EM propagating waves. Why? Because there is not sufficient tension on the Plenum to actuate articulation. The Plenum has to reach a certain critical degree of tension (tension is rotation) for the Plenum to break-up (so to speak) into units; units are nature's way to quantify the Plenum, which is to say going from a homogenous medium to one that is divided up into quantum units.

Waves pass through each other the way ocean waves pass through each other without a loss of energy. Space is full of propagating EM energy of all frequencies. Until this energy finds a way to interact three-dimensionally, matter will not form. Stars provide the environment where EM energy can interact, that is because stars are where three-dimensional interactions take place naturally. Stars are also where the Plenum gets sufficiently tensed as to cause the Plenum to articulate as the model in Fig. 1.25, 1.26, 1.27 demonstrates. When three such planes of the Plenum at the gamma frequency interact, the tension in the Plenum quantizes. That is what is happening in stars. In super nova, the twisted Plenum becomes severe enough to rip mass apart and relieve the Plenum's tension causing the release of a huge burst of radiant energy. The twist is removed from the Plenum, the gravity associated with that twist disappears, and the Plenum is relieved of it stress.

Nature creates from EME (electromagnetic energy) the lambda first and then lets the decay products of a proton, electron and neutrino, bind into hydrogen without the need of high-energy particle physics. The decay products of the lambda are the mass numbers known to physics when measured in units of electron mass. **The numbers are the same numbers obtained by a count of the lines in a STF (Sierpinski Triangle Fractal). The table below establishes a direct and hard connection between the Standard Model using electron mass units and the field-structures model using loop counts.**

+ 2187	=	lambda	the seventh iteration of the STF
- 243	=	meson	the fifth iteration of the STF, (a muon)
- 81	=	electron	the fourth iteration of the STF (beta-electron)
- 27	=	tau-neutrino	the third iteration of the STF
1836	=	proton mass in electron units	

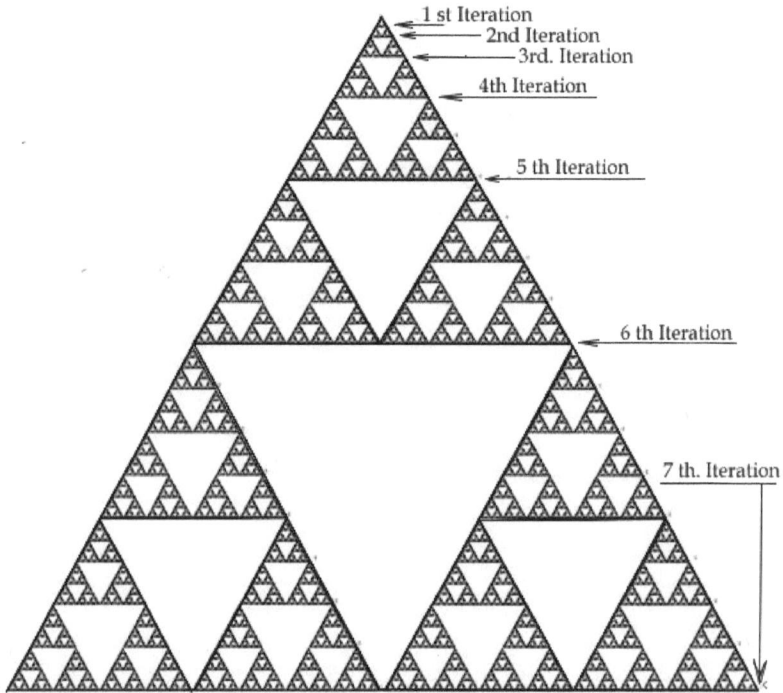

Fig. 1.22 – The Lambda, the full 2187 lines of a seven iteration STF

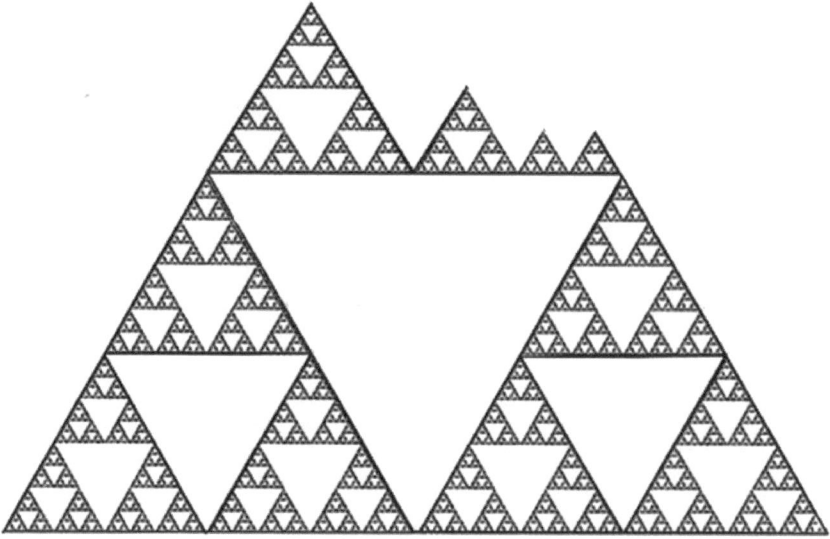

Fig. 1.23 – The Proton, having 1836 loops.

This "mountain range" appearance in Figure 1.23 above shows the 1836 loops that are retained by the proton from the original 2187 loops of the lambda after the lambda has decayed.

> Standard Model physics explains that there are two "Up quarks" (u + u) and one "Down quark" (d) needed to form a proton.

> Unanswered has been the question, "Why would quarks have different energy values"?

> Up quarks have a valance energy of 4.8 MeV (with a 0.05% + or - margin of error)
> Down quarks have a valance energy of 2.7 Mev (with a 0.05% + or - margin of error)

FST explains that the two Up quarks are the two bottom triangles in the STF, each having 729 loops and the remaining triangles above make up the one Down quark (above) with 378 loops.

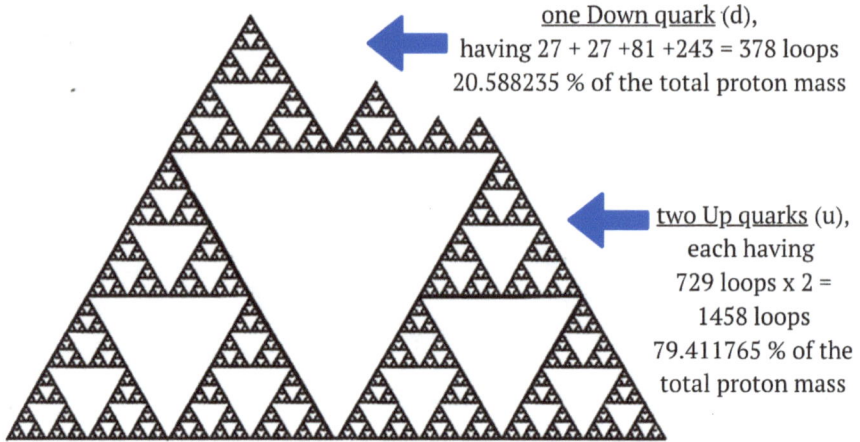

one Down quark (d), having 27 + 27 +81 +243 = 378 loops 20.588235 % of the total proton mass

two Up quarks (u), each having 729 loops x 2 = 1458 loops 79.411765 % of the total proton mass

Fig. 1.24– This diagram explains why quark proton loops have two loops of the same energy and one loop of another lesser energy?

An *Up* quark (u) has a mass of 2.7 MeV/ c² with a margin of error at +.05 to -4%. <en.m.Wikipedia.org>
Considering the margin of error, the mass of an *Up* quark could be as low as–.04% of 2.7 MeV/c² = - 0.148 MeV/ c²
Thus 2.7 – 0.148 = 2.552 MeV/ c²

The *Down* quark (d) has a mass of 4.8 with a margin of error at +.05 to -3%.
  <http://phys.org>
    +0.05% of 4.8 = 0.0625
    0.0625 + 4.8 = 4.8625 MeV/ c²
    4.8625 +0.0625 = 4.925

2.552 MeV/ c^2 divided by 4.925 MeV/ c^2 = 0.51817
Standard Model = 2.552 MeV (u) divided by 4.935 MeV/ c^2 (d) =
0.518 MeV/ c^2
FST model = 378 loops divided by 729 loops = 0.518 MeV/ c^2

The ratio between the *Up* quark and the *Down* quark when measured in MeV/ c^2 and when measured in loops is the same.

Importantly, Fig. 1.24 shows why quarks have different energies, which is to say, they have different loop counts. This shows measuring in MeV, or in loops, give the same ratios and thus produce the same results even though the methods of measuring are completely different. The advantage of using loops is that it validates the use of Tometry's loop analytics and the Field Structure Theory's structural models. It allows for the use of whole numbers to calculate mass values. Importantly, the STF shows why action (energy) come in loops that can have different energies.

7 - The Question of Symmetry

The 1836 loops of Fig. 1.24, shown above, do not have symmetry that we like to see in physics. The STF should not be understood as an actual physical structure. Rather, it is a diagram. It is a mathematical interpretation of how energy accumulates that happens to organize as nature organizes. The structural interpretation of the STF can be seen in the form of a fieldstructures such as shown in Fig. 1.19. To see how the STF diagram translates into a natural world of the Tometry analytic, it will be necessary to transpose the idea of the STF into the isometric triangular matrix of the Plenum thus allowing an understanding how particle formation affects the Plenum.

NOTE: *The illustrations above are two-dimensional. A three-dimensional form of the STF would give a fuller presentation, but in a paper presentation, the depiction would be visually difficult to understand.*

The removal of triangle iterations from Fig. 1.23 to get to Fig. 1.24 reveals what would appear to be an asymmetrical proton. That is not in keeping with nature's preference to have things symmetrical. By translating the STF to an isometric triangular matrix reveals the proton's symmetry. (Fig. 1.28). It also reveals the structure of the Plenum.

Before delving into the isotropic triangular matrix of the Plenum (Fig. 1.28) it will be useful to point out something special about the Plenum. A question often asked about the idea of a Plenum is how can an absolute non-temporal, non-spatial, event-less medium such as is the Plenum become full of events, fully delineated, fully divisional, fully articulated and then suddenly become quantified having space/time attributes? Figures 1.25, 1.26, 1.27 show how this astonishing feat is accomplished.

Though the Plenum in its un-modified state is non-dimensional, when stressed by twisting, develops an isometric triangular geometry instantly throughout its entire extent. Figure 1.26 shows how a tensed homogenous strip of metal having no internal divisions, when twisted will at a certain point break-up into a triangular pattern over its entire surface. The Plenum in its full entropic state is like a plain piece of metal (Fig. 1.25). It has no division, no matrix, no articulation. It is homogenous. When tensed and then twisted, the metal will take a certain degree of twist before reaching the critical amount of twist, after which a geometry suddenly appears instantly over its entire length. The tension quantizes the Plenum, not intermittently, not locally, but over its entire surface, virtually instantaneously.

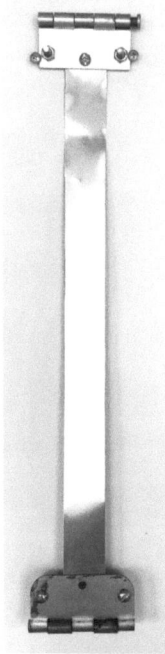

Fig. 1.25 - strip of metal not twisted.	Fig. 1.26 – metal strip at the start of being twisted. Without tension, the metal simply curves and remain continuous.	Fig. 1.27 – metal strip being tensed and twisted suddenly breaks into triangles.

To understand why EM gamma radiation will spontaneously flip into a particle, this model is useful. When three loops tensing the Plenum to gamma ray frequency, causes the Plenum to quantize by flipping into a closed bound matrix and reveal its polyhedral geometry, a fieldstructure is born. A full study of this interesting mechanism has been explained in another paper. Useful for this discussion, what is seen in the experiment above, is to show the mechanism for how the non-local Plenum can become an articulated, quantized wave having frequency.

It is when the Plenum is condensed through twisting that localized matter forms. The mechanism that transforms the matrix

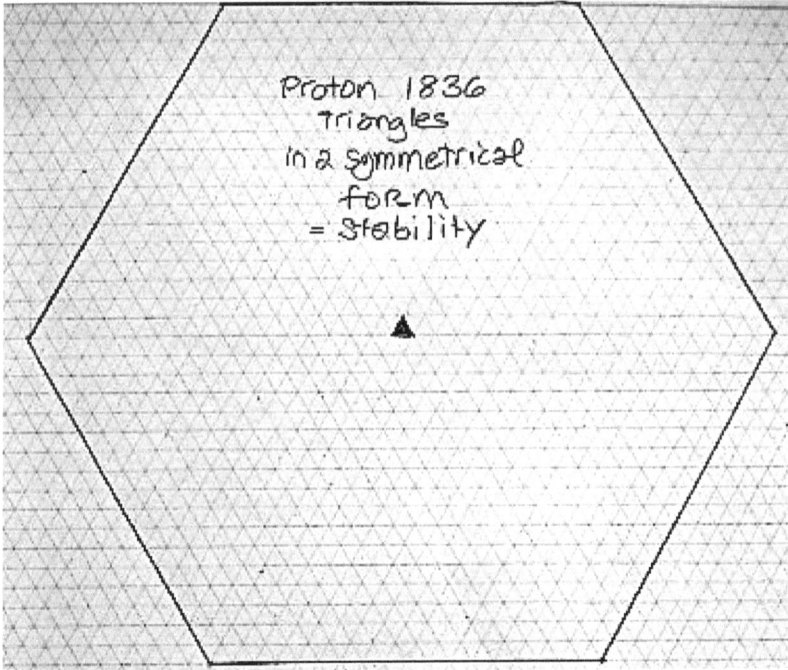

Fig. 1.28 - 1836 triangles (loops) arranged around a single black triangle in the center of the matrix producing a semi-symmetrical outside border with the two-fold symmetry of a hexagon. This hexagon is semi-symmetrical having two more triangles on three of its six sides. Having semi-symmetry is not an imperfection. This form allows the real-matter wave and the anti-matter wave to interact in such a way the deficiency of one side of the hexagon, the two sides missing two triangle units, is infilled with the side of the anti-matter wave. In that way symmetry is achieved. It takes the real and anti-real worlds of form and structure to achieve symmetry.

from a non-local homogenous continuum to localize asymmetrical chiral particles is tension. For ease of viewing in this paper, the Plenum is shown two-dimensionally in Figure 1.28. This Plenum diagram is useful to see why the proton is a combinatory particle. It does not convey the symmetry which is essential for a stable particles such as a proton. This symmetry is necessary for stability. Symmetry is another way of saying equilibrium. As has been stated previously, nature condenses loops and entangles them into a particular form that can equalize the energy throughout the structure. Particle formation is accomplished by condensing the Plenum's loop. Atom formation is accomplished by decaying a condensed lambda particle/field which then deploys into hydrogen. The Periodic Table of Elements is built on the prototype hydrogen particle.

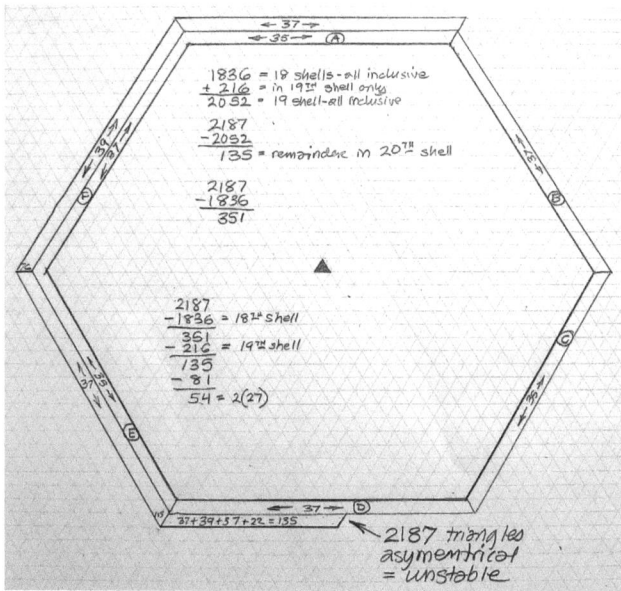

Fig. 1.29 - Shows that the lambda does not have a symmetrical field form. Adding loops outside the hexagon of 1836 loops of the proton produces a 2187 lambda particle that will be unstable.

To portray the way the Plenum creates particles, each triangle in Fig. 1.28 is a loop unit of the Plenum. Mass forms when a sufficient number of triangles are collected on to one triangle at the center of the field that will be the nucleus of the mass event.

There are only two symmetrical configurations of triangles shown in Fig.1.28, that of the single triangle having three loops (in black) and the proton having 1836 loops in the quasi- hexagon form. All other iterations of the STF having 9, 27, 81, 243, 729, and

2187 are not symmetrical energy patterns of an isotopic triangular matrix and will self-destruct.

The charge field surrounding the dark triangle proton is the area of the Plenum into which the surrounding 1836 triangles have condensed to make the proton.

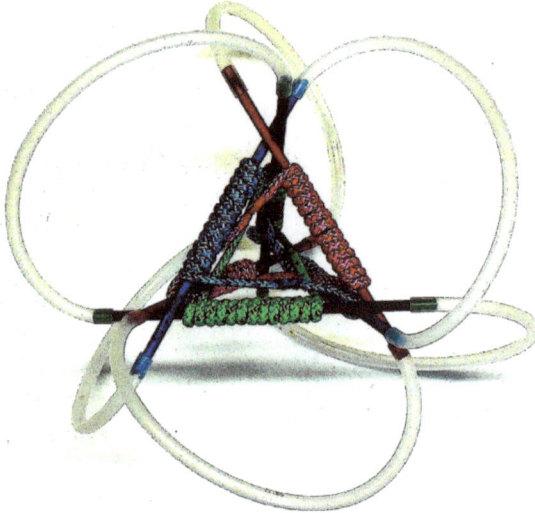

Fig. 1.30 - A Wysocki/Briddell- Fieldstructure (WB-FS) with one side of the wave-set (the dark strings) having been condensed to the nuclear polyhedron leaving the other side of the wave-set (white tubes) deployed. One side of the wave (colored strings) condensed to the nucleus. The wave is constrained to the nucleus by the three-dimensional magnetic forces of an electromagnetic charge.

8 - Fractional Values of Particles

A full accounting of the electron has to include the electro-neutrino mass found in the electron's deployed action field. In Fig. 1.30, a particle has a nucleus (the polyhedron at the center of the form) and an action field that extends out from the nucleus. As has been shown, the formation of the nucleus involves the condensation of one side of the wave-set which takes the energy of the entire wave-set to the nucleus, except for one unit of twist that remains in the deployed wave. Because the wave is a loop it will always have one unit of twist.

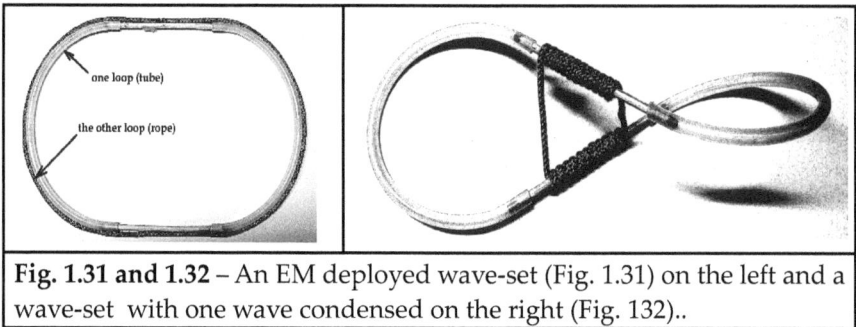

one loop (tube)

the other loop (rope)

Fig. 1.31 and 1.32 – An EM deployed wave-set (Fig. 1.31) on the left and a wave-set with one wave condensed on the right (Fig. 132)..

Shown above is how one side of the wave condenses (the dark string), while the other wave (white tube) remains deployed. Note that when one side of the wave-set condenses, the other side of the wave-set is forced to bend. Bending puts tensional energy into field. Three of these condensed wave-sets form Fig. 1.30.

9 - Particle Hierarchy

- The action field particle of the proton is the electron.
- The action field particle of the electron is the electron-neutrino.

The mass in the charge field of a particle, as well as of an atom, is attached to the nuclear mass of the particle (as shown in Fig. 1.30). In the case of the electron, the neutrino's mass in the three charge field loops is tiny. Nonetheless, the neutrino's mass while tiny adds up as the fractal *uperates* (iterations that get larger) and has to be factored into the calculation to arrive at the total mass in any particle. The only way to get whole number values without a fraction is to begin with a whole number unit of measurement that is sufficiently small to be truly fundamental. Present science has discerned that the ultimately fundamental particle, which does not seem to have a *downerated* fractal supporting it, is the *Planck particle*. In FST terms the Planck particle is called the **Plenum Particle**. Starting with the Plenum particle, the particle hierarchy of mass can be built with whole number values, without fractional values, by following the STF iteration of masses. As of this writing I do not know how many iteration sequences there are between the Plenum particle and the electron particle. For the sake of brevity, I have chosen to begin counting particle masses with the Neutrino Sequence which lives in a space significantly larger than a Planck length. The neutrino scale of a particle is within reach of present science.

The role played by Magnetic and Electric forces

In real-matter, the **electric force** is generated by the condensed left-handed, clockwise rotating, wave/loop. The **magnetic force** is generated by the deployed right-handed, counter-clockwise wave/loop. This is why the magnetic force is a field force and electrons are a field object force, one is perpendicular to the other. Fieldstructures explain why the magnetic field can induce movement to the electron and inversely why the electron motion can carry a magnetic field with it as it moves. The magnetic force condenses the electron field with each loop of the magnetic field. The two forces have to balance. If one is not balanced by the other, the forces move to readjust. Looping produces a magnetic force and the magnetic forces condenses the electron wave proportionally. If the magnetic field is twisted by rotation of the field, then the electric field has to move to balance the magnetic field's twist.

The problem with using the neutrino as the field particle of the electron has been that the mass of the neutrino has not been precisely known. Complicating the use of the neutrino as a measuring tool is the fact the neutrino comes in three different energy forms. Physics has been summing the three mass values and using that as its mass number. The mass number of the neutrino has been found experimentally, not theoretically. Only theory can reveal the internal mechanics of a form.

10 - Neutrino

Field Structure Theory (FST) and Tometry have found a way to obtain the exact value of the three neutrino forms. FST has done this theoretically and confirmed it using the values found experimentally and integrating them into Field Structure Theory using the Tometry analytic. Particle hierarchy is organized fractally by iterating down from the known value of the proton. An empirical measurement is not needed once the structural system of nature is known. Knowing the structure of the fractal hierarchy allows the known values to predict the unknown values that are experimentally out of reach. This allows the analytic to go where experiment cannot.

Physicists Mainz and Troitsk have stated, *"Currently the best kinematic upper limits on the neutrino mass of 2.2eV (0.0000022 MeV) has been set by experiments, using tritium as a beta emitter."*

<https://en.osti.gov/biblio11867-absolutENeutrino-mass-measuremENts>

Using the Mainz and Troitsk experiment numbers for the collective mass of the electron-neutrino (at 0.0000022 MeV) makes the electron 189,319.674+ times more massive than electron-neutrino when *uperated* seven times. <https://aip.scitation.org/doi/abs/10.1063/1.1470264?journalCode=apc>

Using the 0.0000022 MeV number of Mainz and Troitsk for the mass of the neutrino as acceptable, the mass of the proton at 938.27231 MeV makes the proton **347,619,912**.041304851538595 times more massive than the neutrino. The mass of the lambda is 1115.683 MeV making it **414, 547,375**.327445211901143 times more massive than the neutrino. The difference in neutrino units between the proton and lambda is **66,927,463**.2861. These **66,927,463**.2861 loop masses are the meson, electron and neutrino decay products of lambda. **414, 547,375 minus 347,619,912 = 66,927,463.**

Field Structure Theory with the aid of the Sierpinski Triangle Fractal (FST), downerates (iterate down) from the known mass values of the proton and electron to find the mass of the neutrino. If it is correct to include the neutrino into the mass of the electron, as FST maintains, it means that the electron has internal structure. Presently, as far as I can tell, physics does not consider that to be the case. The reason neutrino particles that make up the electron don't show up in electron/positron annihilation is because the loops once broken revert to being EM (x-rays) waves and not particles waves. A neutrino is a "particle wave".

Returning to look at how loops calculate the mass of the electron-neutrino, FST puts these numbers from physics into the context of loops. The fractional value of the proton when measured in electron masses is 0.152672. Each loop carries a third of this number which is 0.05089081666666>.

Reducing this 0.05089081666666> fraction by seven iterations of the STF (729), gives the mass value of 0.000069809+ for neutrinos in the electron's action field. [Note *that I am referring the mass of the charge field particle, not the mass of the electron's nuclear field.*] There are three neutrinos in the electron's charge field. Each of those three neutrinos are tau-neutrinos. To find the mass value of the simplest neutrino is to find the value of a single electron-neutrino.

To get that number, the tau-neutrino has to downerate to the base unit of the neutrino family which is going from 729 loops to 3 loops. 0.000069809+ downerated seven times is **0.00000028728015.** This is the value in MeV of the electron-neutrino according to FST.

The nucleus of any fermion is composed of three condensed left-handed clockwise spinning EM real-matter loops and three right-handed counter-clockwise anti-matter loops. The three right-handed EM loops are deployed in the charge field and are electron spark loops. The electron deployed loop is right-handed and is partnered with a positron condensed loop that in the condensed state is the proton. The positron loop condenses 100,000 times in volume when it becomes a proton. Inversely, in the anti-matter side of the universe, condensing the electron's loop 100,000 times, while leaving the positron loop deployed, transforms the electron into an anti-proton.

> **The Plenum pervades everything and hence ties together matter in a web of loop energy. The implication is that all masses are various degrees of condensed EM loops, and EM loops are deployments of the Plenum.**

The reason this has not been apparent is that particles are not the product of a single loop/wave. It is by loop interaction of three loops that matter forms and it is by loop interaction, which is an EM interaction, that loops condense and deploy. Three interacting wave-set loops are the minimum number of loops needed to a stabilize energy three-dimensionally into a form having mass. As the loops condense the particles associated with these condensing waves take on unique statistics that have seemed previously to be empirically separate and unconnected. At the level of fundamental particles, the moment of transition is so fast, as if instantaneous.

Interactions at that speed have not been observed. Field Structure Theory, because it can be modeled in the macroscale, shows the start-state and end-states of a transition from one form to another. How the form changes can be observed at the macroscale in smoke rings. FST proposes that it can be seen kinetically in the macroscale when done with electron rings. How these seemingly instantaneous transformations occur is the key to unlocking the mysteries of particle formation. Particles are end-states of a transformational process and in themselves without a proper model, do not give us a clue as to what transpires during the interaction process.

The pervasiveness of EM loops and how they deploy and condense can be seen in particle/anti-particle annihilation, wherein bounded particles becomes radiant EM waves; when electron/positron annihilation occurs, two chiral gamma rays result. Physics needs to shift from being a study of parts (particles) to the study of wholistic field transformations, i.e., energy transformations. It is what goes on between start and end-state particles that is important. Field mechanics are what link together start and end-states.

The value of an electron-neutrino according to FST has been established to be: 0.00000028728015 MeV.

This number closely agrees with the estimated mass number provided by experimental physics. The Mainz and Troitsk experiment weighed-in the electron-neutrino at 0.00000022. The FST findings differ by 0.000000067, close enough to be considered a match.

11 - Energy Spectrum of First-Generation Particles

Listed are the energy values associated with the neutrino and electron sequences. There are energy states below the electron-neutrino, but their energy/masses are diminishingly small, at least a billion times less energy than the electron. Without considering that there are particle sequences below the energy value of an electron-neutrino, the following spectrum of energy of all first-generation particles listed by the Standard Model of Particle Physics, can be determined. (Note: The Standard Model's first-generation is FST's second- generation.)

Neutrino sequence begins in loops with 1 electron-neutrino and ends with 726 loops in the tau-neutrino. It takes three tau-neutrinos with a combined mass of 2187 (three times 729) to make an electron. The electron sequence begins with 2187 electron-neutrino loops and ends with the lambda hyperon having 1,594,323 electron-neutrino loops. If there are discovered particles smaller in mass than the electron-neutrino, they would be in a sequence below the neutrino's sequence and would add mass that is a billion times smaller to the neutrino.

12 - Electron's Relationship to Atomic Structure

List #	Shell #	electrons possible in shell	Shell Letter name	Loops In Shell	energy Totals in MeV	Particle Associated With each shell
1	1	2	K	3	0.511	electron
2	2	8	L	9	4.599	electron
3	3	18	M	27	13.797	electron
4	4	32	N	81	41.391	Electron (Beta)
5	5	50	O	243	124.173	meson
6	6	72	P	729	372.519	koan-meson
7	7	98	Q	1836	938.196	proton
8	NA	NA	NA	2187	1115.683	lambda

Table 1 – Electromagnetic energy and the electron shells to which these energy numbers apply. The ratio of values between loops matches the ratio values between shells in MeV. The MeV values are listed at:
<Pinterest.com> and <EN.mwikipedia.org>

The electron in the atom can have a maximum energy up to 81 times greater than its rest energy. That is why there are only 81 stable elements. Beyond the 81 stable elements, the electrons have too much kinetic energy to remain attached electromagnetically to a proton. Each iteration of the STF and the various combinations of iterations are the energy found in the shells of the atom.

The ratio between the various iterations of loops in Table 1 and the ratio between the MeV values of various particles are the same, indicating that counting loops in the STF is a valid and verifiable way of explaining the structure behind mass and energy values. Energy when bound up in atomic structure organizes as shells. There are seven shells that can be found in atoms; there is no eighth shell.

The seven shells possible for an atom relate exactly to the seven energy loop iterations of the Sierpinski Triangle Fractal (STF).

Beyond seven shells the energy of an electron is too powerful to be held by the electromagnetic attraction of the nucleus. There are 92 natural elements. 81 are stable and 11 are unstable, i.e., radioactive.

Atomic Shell	energy in MeV	energy in loops	Iterations in multiples of 3	Name of particle in Standard Model
7th shell =	1117.557 MeV	= 2187 loops		[lambda hyperon]
6th shell =	372.519 MeV	= 729 loops	(729 x 3 = 2187)	[tau]
5th shell =	124.173 MeV	= 243 loops	(243 x 3 = 729)	[meson]
4th shell =	41.173 MeV	= 81 loops	(81 x 3 = 243)	[tau-electron]
3rd shell =	13.797 MeV	= 27 loops	(27 x 3 = 81)	[meson-electron]
2nd shell =	4.599 MeV	= 9 loops	(9 x 3 = 27)	[electron
1st shell =	0.511 MeV to 1.533 MeV			
0.511 MeV x 3 = 1.533 MeV = one loop (0.511) to three loops (1.533)				
1.533 MeV x 3 = 4.599 = three loops (1.533) to nine loops (4.599) and so, on.				

Table 2 – Comparing ratios of loop counts to MeV energy, showing congruences. Should the amount of energy of each shell condense into a particle(s), the particle's name is given. The MeV numbers and the particles associated with a particular number are the known values. <*EN.m.wikipedia.org*>

I have found several values of the lambda hyperon in physics literature, 1117.557 & 1115.683. Below I look at how the 1117.557 MeV number parallels the loop counts of seven iterations of the SFT. As can be seen the numbers are very close. On closer analysis it may be that the field particle energies are not being counted, in the same way the mass of the electron has not included the mass of the electron-neutrino field particle.

1. 1117.337 (the lambda mass) minus a meson at 124.173 MeV = 993.164 MeV.
2. The mass of 994 is given for the positively charged koan-meson. Remove from the 993.164 koan-meson two 27 electron masses and neutrino mass of 1.533 equals 937.631.
3. The mass of the proton shown in Table 3 is 938 MeV (a rounded number).
4. The fractional number difference between 994 and the 993.164 numbers comes from the fact that the contribution of the electron-neutrino masses have not been included.
5. As seen above, the effect of adding loops in multiples of 3, increases the energy of the particles at the same ratio as the MeV numbers increase, which are iterations of the STF.
6. The energy/mass of the photon and photon wave-set can be anything since they are not as yet interacting with other loops in such a way as to produce a Fieldstructure.
7. It is only by interacting loops using Tometry analytics will the known values of energy/mass be produced.
8. Fieldstructures account for the appearance of forms having locality, mass and duration.
9. Mass occurs when there is sufficient loop/twisting (energy) to condense the loops from one state to another in "jumps" from 1 to 3 to 9 to 27 to 81 to 243 to 729 to 2187, or in any combination of those numbers.

10. Fermion energy increases by quantum jumps of 3. Boson energy can increase arithmetically (1,2,3,4,etc).

Mass numbers increase by multiples of three loops.
Energy numbers increase by the square of the mass number.

Said differently, the total mass and energy numbers increase by the cube of the mass number. $E = L^3$ (energy = loops cubed)

Mass sequence	Energy=twists on loops	Particle and shell iteration
Loops agree w' data	Numbers agree w' data	
1 loop	$1^2 = 1$	electron – 1st iteration (shell)
3 loops	$3^2 = 9$	electron – 2nd iteration (shell)
9 loops	$9^2 = 81$	electron – 3rd iteration (shell)
27 loops	$27^2 = 729$	electron – 4th iteration (shell)
81 loops	$81^2 = 6,561$	electron – 5th iteration (shell)
243 loops	$243^2 = 59,049$	muon electron – 6th iteration (shell)
729 loops	$729^2 = 531,441$	tau - electron – 7th iteration (shell)
2187 loops	$2,187^2 = 4,782,969$	lambda – maximum in 1st generation

Table 3 – Mass and energy loop sequence as related to electron shells and the associated iterations diagrammed by the Sierpinski Triangle Fractal.

The difference between the numbers physics has for mass and energy values versus the numbers Tometry gets is so slight as to suggest correlation. In other words, by using loop and twist counts, Tometry gives virtually the same number as those numbers arrived at experimentally in MeV. This means that with Tometry there is a way to explain mass and energy values of particles, and at the same time have a Tometry (geometry) that can model the field form empirically at the macroscale. This is something physics has long needed.

To the theorists, knowing why particles form hierarchy using Tometry is good. For those who apply theory to practical application, knowing how to do this technologically is even better.

In the lambda decay where a proton (p+), meson, electron (e-) and neutrino are created, hydrogen does not form because the e- is a highly kinetic (hot) beta particle. The proton's positive charge is not strong enough to hold the e- that would permit the hydrogen atom to form. If the e- were cooled sufficiently, the e- would be able to bind to the proton and become hydrogen. Where in the universe, one might ask, can a cold lambda particle form from EME energy and then immediately decay into a hydrogen atom? In other words, where in the universe can gamma EME find a super cold area to interact three-dimensionally and condense into a lambda particle? The lambda, that is known to physics, is a particle created by high-energy collision. Hydrogen, however, is not created in the laboratory where lambda energies are extreme. Here I am suggesting that there are places where a lambda can be created from EME. Those places are where EME can condense in a super cold environment. The cold decay of a lambda particle that will form stable hydrogen will not be in a hot collision environment, hence this mechanism has remained undiscovered.

Physics assures us that protons electrons and neutrons were all made at once in the intensely hot Big Bang (BB). In the BB scenario, the stuff of the universe then floated in turmoil cooling as it dispersed until the electromagnetic force could have enough force to make particles. This was followed by the gravitational force gathering particles into LMOs (Large Massive Objects), i.e., the stars. That scenario in itself is so full of fallacies, that the miracle is anyone should believe in this cosmology. The biggest problem of all is how this BB theory has to violate all the laws of physics to make it happen. It also violates all the laws of reason. FST finds the BB scenario preposterous. Physics is saying we can never know how the universe works, because what caused the BB in unknowable. If you conclude how the universe started is unknowable, you have surrendered all hope of ever knowing how the universe works. This is a product of the "something from nothing" creation myth shared by science and some religions. FST does not ask you to surrender your belief in reason and succumb to fantasy.

FST, meanwhile, postulates the manufacturing of mass and energy without violating present time physics.

FST ask what if you had a mechanism for creating mass that did not need to be supported by the *something from nothing* conjecture? What if the production of mass was simply an ongoing process fueled by mass deploying into energy and energy condensing into mass, cycling endless?

From the center of a galaxy there is a constant stream of EM energy spiraling out perpendicularly from the galaxy's plane of rotation. It is only visible if there is matter (dust) above the galaxy's plane of rotation that becomes illuminated by the plum of EM energy. This plume of gamma energy is the result of mass being torn apart by the extreme motion of the vortex at the center of a galaxy. The center of the galaxy is void of matter. It is not a black hole filled with compressed mass.

The opposite process of producing particles from this plume of gamma radiation is the work of stars. EM energy produced at the galaxy's core, flows back into the galaxy's stars. Streaming into the star at its poles, this EM energy finds its way to the star's matter-less cold core, condenses into lambda particles, which then decays into the protons, electrons and neutrinos to form atoms of hydrogen! Being a closed loop that goes from energy to mass and mass to energy cyclically, is a far more satisfying than a Big Bang that began arbitrarily and self-destructs in the end. What a spectacularly demoralizing conclusion.

Table 4 shows the lambda decay sequence in MeV and compares that to the same decay sequence using loops as organized by the STF.

Table 4 – Lambda -Proton relationship
Proton mass = 938.2720813 MeV
Lambda mass = 1115.683 MeV

1115.683 - lambda in MeV
-938.272 - proton in MeV
 177.411 - difference between lambda and proton mass in MeV

177.411 divided by 1115.683 = **0.1590155977>** percent
difference between lambda and proton mass in MeV

 2187 - lambda mass in loops
-1836 - proton mass in loops
  351 - difference between lambda and proton mass in loops

351 divided by 2187 = **0.160493827>** percent difference
between lambda and proton in loop mass count

0.160493827
0.159015598
0.001478229 - difference between the Standard Model using
               MeV and Field Structure Theory using loops.

This chart shows how similar to each other are the Standard Model using electron mass values and the Field Structure Model using loop counts.

13 - Summary of the Fieldstructure Hierarchy

The fieldstructure way of modelling particles explains fermions in terms of loops and twists of those loops.

Deployment and condensation are the two processes in nature that work together to produce fundamental form and structure from EM waves.
Deployment is a condensed wave/loop that is, or has been, allowed to deploy its frequency over the wave-set to which it belongs.

When full deployment is reached, each side of the wave-set has the same frequency, occupy the same volume and will be in a state of equilibrium. Charge disappears in a wave-set that has no longer separated its deployed and condensed loops. Charge only occurs when the wave-set has had one of the two loops condensed. It is that condensed loop that now has charge.

Fig. 1.33 – A deployed wave-set is comprised of an equally deployed right-handed counter-clockwise wave/loop (red line on loop) and left-handed clockwise wave/loop (black line on loop). While I assigned separate handedness to each loop, the red and black loop designation, when the loops are equally deployed as shown in Fig. 1.33 there is no way of telling the handedness of the loops. Handedness (and charge) only becomes apparent when one or the other loop condenses leaving the remaining loop deployed.

Condensement is when one wave of a wave-set has condensed to a locality on the deployed wave, causing an asymmetry and a dis-equilibrium. Condensement makes it possible for non-local energy to become localized and available for electrical use. An electric current is a condensed EM wave that is available for redeploying back over the charge-less deployed circuit as electrical energy. In nature, the condensement occurs in the middle of the deployed wave as shown in Fig. 1.35 and not as shown in Fig. 1.34.

Fig. 1.34 – A condensed wave-set is a wave-set comprised of a right-handed wave/loop (red line on loop) that is deployed and a left-handed wave/loop (black line on loop) that is condensed. The condensed wave will deploy back over the deployed wave unless confined. Note how the condensed wave (black line) has taken all the twist in the wave-set with it leaving the deployed wave (red line) with only one twist.

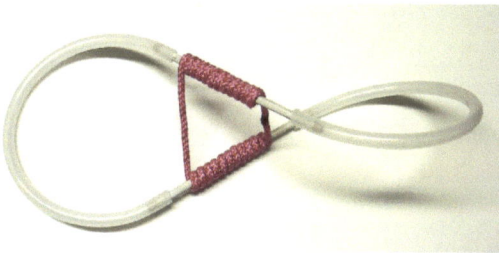

Definition: **Wave-set**
When two chiral waves shown in Fig. 1.35 entwine (entangle), one rotates clockwise (CW) and the other rotates counter-clockwise (CCW).

Fig. 1.35 – Shows how a wave in nature condenses to the middle of the wave-set. In doing this, the deployed wave (white tube) becomes twisted and bends, creating torsion energy in the loop. Why a loop (red string) condenses is not apparent when only a single wave-set is involved. When three EM wave-sets interact, the pinching magnetic forces squeeze which-ever loop is inside and that squeezed loop condenses to form the nucleus. Figure 1.36 shows what happens when the three loops interact electromag-netically. The white tubes are the circuitry of the electron loops, and the red string is the positron loops that have condensed. The resulting structure is in torsion. The deployed loop (plastic tube) is twisted due to condensing of the red string. Twisting in the form gives the loop torsion. Torsion manifests as magnetism. Space is bent and that produces a magnetic field in which the electric force circulates.

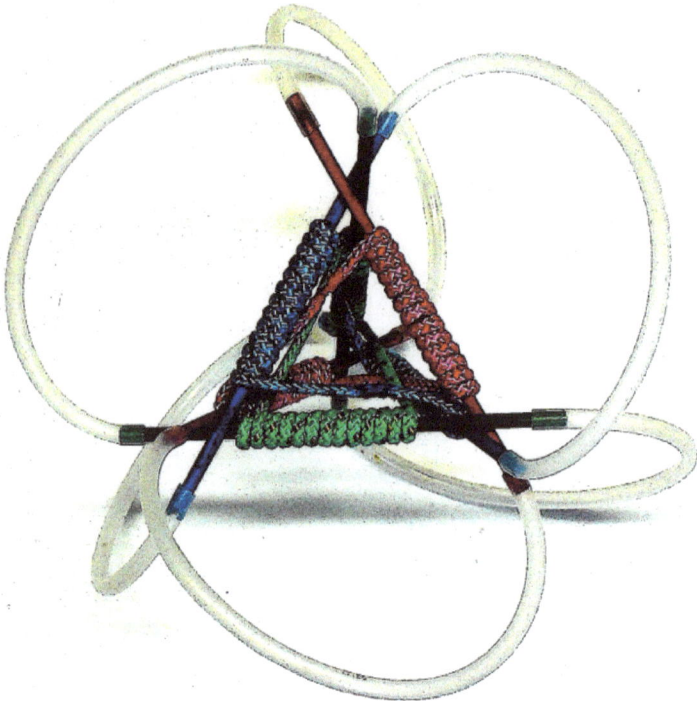

Fig. 1.36 – Tet- Structor (face view) - A tetrahedronal Structor with three wave-sets interacting to produce a nucleated structural form.

In the illustration, each wave-set has 26 loops condensed to the nucleus, half on one side of the loop and the other half on the other side. This can be seen in how each colored string is divided into two clumps of loops on opposite sides of the tetrahedron at 90 degrees angles to each other. The deployed loops (the white tubes) account for 3 loops simply because of being a loop. Adding the 78 condensed nuclear loops and the 3 deployed loops = 81 loops. A lambda particle would factor 2187 loops times the number of loops in the deployed charge field. If there are 3 loops in the charge field, there would be 3 x 2187 = (6,561) loops in the nucleus and those loops would be twisted 2187^2 (4,782,969) times. In the nucleus of this model is 26 x 3 = 78 loops.

Fig. 1.37 – Vortex view of a Tet-Structor. All particles of the first generation of particles have this structural configuration of loops.

14 - Basic Attributes of Matter in Terms of Loops and Twists

Fermion spin is an affect of loops being bent, compressed and twisted. Spin can be either right-handed or left-handed. All forms of energy, be they boson or fermion have spin. Bosons have spin 1 designation. Both CW (clockwise) and CCW (counter-clockwise) spins are present and share the same domain. Fermions with ½ spin have one loop deployed and one loop condensed so the frequency of each loop is maximal and the other minimal. The *maximal loop* refers to the fact the wave/loop is fully condensed and the *minimal loop* refers to a fully deployed wave/loop devoid of twists. The spin/handedness of a particle decides whether it is a real-particle or an anti-particle. Spin ½ means the presence of only a right, or a left-handed wave, but not both. Spin ½ loops are found in matter. Spin 1 means both right and left-handed wave-loops are present such as found in EME.

Chirality is clockwise and/or counter-clockwise rotating circuitry of a fieldstructure. Forces that are rotating clockwise and counter-clockwise are both condensed and have their chirality in balance and thus have no charge, such as a neutron. Charge is the effort exerted by either a right or left-handed condensed wave-loop to deploy. "No charge" means the energy in the wave is equally distributed between the right and left-handed loops of the wave-set irrespective of whether or not the loops are deployed or condensed. Charge means either the right or left-handed loop is condensed, but not the other loop, which remains deployed. In the real-world of matter, a positive charge means the loop is rotating left-handedly clockwise, such as a proton. Negative charge means the loop is right-handed and rotates counter-clockwise, such as an electron. Density is the number of loops in a volume of space and relates to the mass of a particle.

15 - Real-Matter (RM) and Anti-Matter (AM)

1. Real-matter (RM) and anti-matter (AM) depends on the rotational handedness of the loop and whether the loop is condensed or deployed. Example: A condensed positron loop is RM. A deployed positron loop is AM. Without a model to inspect this is hard to imagine possible, that a

wave (particle/wave) can change from anti-matter (the positron) to real-matter, without changing handedness. There is an easy explanation for this phenomenon; the positron is no longer cloaked by the electron when it condenses.

2. When a left-handed EM loop, called a positron loop, condenses to the nucleus, a proton is formed.

3. When a right-handed EM loop, called an electron loop, condenses to a nucleus, an anti-proton is formed unless it is condensing onto a proton which would make the proton into a neutron.

4. However, if there is an extra proton in the nucleus, an electron loop can condense to the nucleus to create a neutron with a neutral charge and establish electromagnetic balance.

5. Anti-matter (AM) interacting with real-matter (RM) in the same wave-set is why waves wave.

6. A wave-set is entangled and is structurally stable.

7. The AM is not visible and cloaks the RM as the waves wrap around each other helically, causing the RM to appear to pulsate from maximum intensity to no intensity, only to re-emerge and become maximally intense once again. The process repeats in cycles measured in Hertz. This is why waves wave.

8. There is equal amounts of RM and AM in the universe.

9. RM and AM are inseparable.

10. RM and AM each have equal but opposite gravity force.

11. RM's arrow of time (present to future) goes in the opposite directions to AM's arrow of time (present to past).

12. Universally, RM's mass equals AM's mass.

13. RM is condensed right-handed counter-clockwise (EME) electromagnetic energy.

14. AM is condensed left-handed clockwise (EME) electromagnetic energy.

15. RM and AM when in the same quantum system are compatible. They share the same axis.
16. When RM and AM are in different quantum system they do not share the same axis of rotation and should they meet, revert to electromagnetic energy and become radiant.
17. Radiant EME returns energy to the Plenum and in so doing removes gravity from the universe.
18. Deploying EME lessen (eliminates) mass and removes gravity.
19. Condensing EME creates mass and gravity.
20. Condensed EME takes Plenum energy and knots it into EME mass structures.
21. Space and time shrink when EME condenses and expand when EME deploys.
22. Space and time are the result of the formation of mass and energy.
23. Thus, if the universe is perceived by RM to be expanding, then it can be concluded that mass is deploying by becoming energy.
24. If the universe is perceived by RM to be contracting, then it can be concluded that the universe is predominately condensing energy into mass, that is going from totality to locality.
25. **Energy to mass equals condensing.**
26. **Mass to energy equals deploying.**

16 - The Neutron

A neutron forms when an electron condense onto a proton. This can happen only as long as there is another proton in the nuclear domain, so that the two protons can share the condensed electron. Two protons in the nucleus are necessary to attract an electron into the nucleus where the condensed electron can be shared. The electron ties the two protons together flipping from one to the other. The neutron outside of a nucleus decays, because the kinetic centrifugal force of a condensed electron that forms a neutron is superior to the magnetic binding force that held it to a proton to make it an neutron.

Condensing an electron to the nucleus condenses the volume of the electron to a space 100,000 times smaller. This compacts the nodes of the electron wave thereby increasing its energy to that of the strong force that is associated with the gluon. **A gluon is a condensed electron.** The gluon cycles between the protons. Whichever one it is near becomes a neutron. A neutron doesn't form hydrogen when it decays, because when a neutron decays, the energy of the departing electron is too strong to be held by the EM force that exists between the proton and electron. A speculation: If a neutron decays at or near absolute zero temperature, hydrogen may form. The decayed electron goes flying off preventing the formation of the hydrogen atom. For the electron to be held to a proton requires the decay to be a cold process; a process previously described, as when the lambda hyperon decays in a cold environment, forming hydrogen.

17 - Loop Condensement

There are five levels of condensement. At present, physics recognizes the quark but not the fact it is a condensed EM loop. Besides the quark loops there are other degrees of loop condensement that form other particles. Not presently recognized, there are five stages of loop condensement between the Plenum and the proton. The names are given in Tometry. After listing these (below), explanations will follow giving the physical manifestation of these Tometry forms.

1. Aum
2. Ark
3. Lark
4. Spark
5. Quark

1. **Aum– The "Plenum", the mother of all loops**

Plenum is background-dependent aether, meaning all mass is derived from the Plenum and remains inseparable from the Plenum. The word aether (ether) is not used in FST. Ether refers to a background independent substrate. As the parent of all other loop families, it is called *Aum*. This word *Aum* was first identified and used properly in human history by Sanskrit speaking Vedic peoples of India. The word *Plenum* is Greek, one meaning of which is *"a state of fullness, a great quantity of something"*. The meaning ascribed to it in FST is *the source of phenomena*.

The various forms of matter are 3-D knots in the Plenum. The Tometry of the Plenum is a chiral isotropic tetrahedronal three-dimensional matrix. Geometers know that the tetrahedron does not close-back and hence cannot be the form of the Plenum. However, chiral tetrahedrons sharing the same space have a cubic form and thus will close-back (fill space). A righthanded or left-handed tetrahedron by themselves will not close-pack. It takes chiral tetrahedra to fill space. (See Fig. 1.38)

Motion is not a spatial dimension as fourth-dimensional geometry proclaims, but rather the kinetic component of 3-D action. While the Plenum can be thought of as static, action in the Plenum is instantaneous and absolute in nature. Absolute motion is also absolute rest. The Plenum in its deployed state has no locality, its action is an everywhere potential. Until the Plenum interacts with itself, creating chirality and knots, no quantification is possible. Mass arises by knotting the

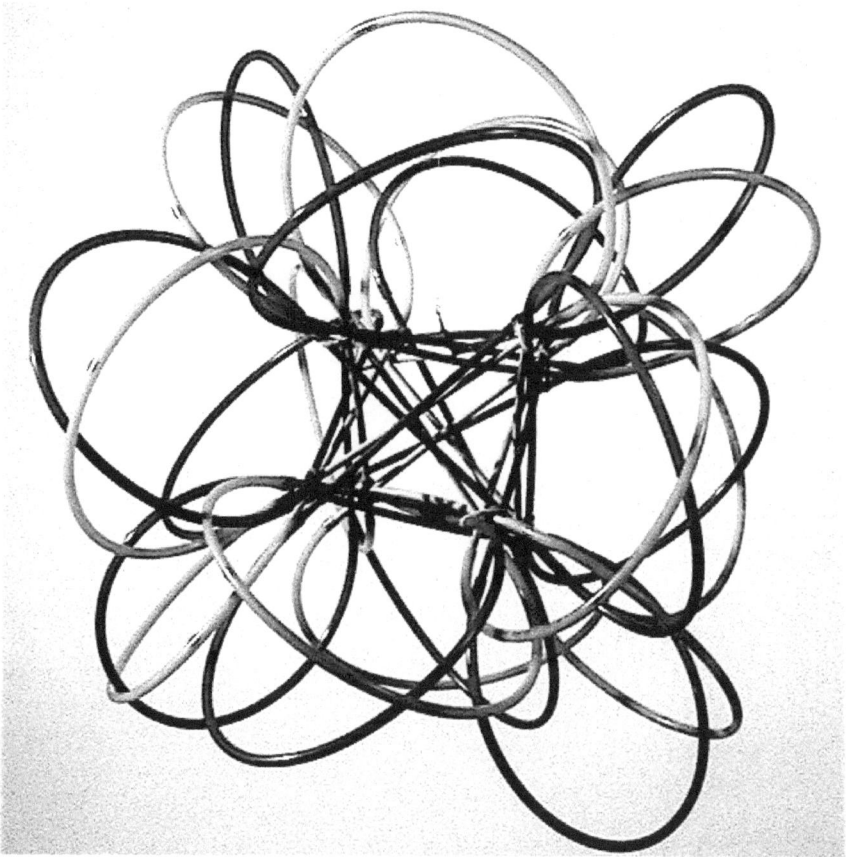

Fig. 1.38 – Chiral tetrahedra sharing the same space, define a cubic form which will close-pack space.

Plenum. When the Plenum knots and forms multiple localities, the localities become relative to each other. However, a locality in itself has the instantaneous quality of the Plenum. A unit of Plenum is a holism as well as the entire Plenum is a holism. Within the boundaries of the localized Plenum, action is instantaneous, although other knots will see it as exhibiting relative motion. A knot has totality. A series of knots linked together also have totality and share the instant in the sense of the action being shared entirely throughout the form. In this way, one instant follows another instant and time is born. The speed of time is a count of how many knots (instants) are linked. Linked knots constitute a quantum system.

2. Ark

The first condensement of the Plenum produces electromagnetic energy (EME). Though Ark loops are energy, they are not particles (defining a limited spatial 3-D domain), they have not as yet compacted to a locality. Ark looping obeys boson statistics. An Ark wave becomes a wave having locality when impacting a mass. when impacting a mass. Once the Ark wave condenses and becomes a locality of energy is then called a "photon particle". To call an Ark wave a particle (photon) is unfortunate in that it leads one to think the wave has locality when a wave should be thought of in non-local terms. It is only when an Ark impacts a mass does the loop behave as a quantum particle. Ark's have two forms, (1) Right-handed, counter-clockwise rotating and (2) Left-handed, clockwise rotating.

3. Lark

Condensed Ark loops produce the Lark loop family. Neutrinos are the Lark family of loops. Neutrinos have spatial domains and obey fermion statistics, except that neutrinos are SOL (speed of light) particles. Being fermions, they are found wherever three Ark loops are interacting, which is tantamount to everywhere. Lark loops are in FST the first-generation of particles and are found in the neutrino sequence. Condensed photons are neutrinos; hence it can be said that neutrinos are condensed light waves, since both are SOL (speed of light) forms of energy.

Lark 1 = one loop = electron-neutrino (0.0000022 MeV)
Lark 2 = two loops = muon-neutrino (0.170 MeV)
Lark 3 = three loops = tau-neutrino (15.5 MeV)

4. Spark

Condensed neutrinos produce the leptons family of fermions. Three lark loops condensed to form a tau-neutrino. Three tau-neutrinos condense to produce an electron.

Spark 1 = electron (at 0.511 MeV)

Spark 2 = muon (at 16.6 MeV) (condensed electron)

Spark 3 = tau (at 177.7 MeV) (condensed muon)

5. Quark

Condensed lepton waves produce hadrons, baryons and hyperons. Quark loops can further condense to produce FST's third and fourth- generation of particles, but decay back to second and first-generation if released from interacting with each other due to the particles having more energy than their structure can contain.

Quark 1 = proton (1st generation particle in Standard Model, 2nd generation in FST)

Up quark = 2 GeV

Down quark = 4.7 GeV

Quark 2 = muon proton (2nd generation particle in Standard Model, 3rd generation in FST)

Strange quark has 92 GeV

Charm quark has 128 GeV

Quark 3 = tau proton (3rd generation particle in Standard Model, 4th generation in FST)

Bottom quark has 418 GeV

Top quark has 1730 GeV

The following chart details how loops condense into the various fields and as a result show how particles associated with those fields form. These loop fields can be modeled by Tometry as three-dimensional Fieldstructure.

Table 5– How EM loops condense into increasingly more massive particles.

Standard Model particle name	How particles are classified in FST
Mass	Energy Source
neutrino	condensed right-handed CCW photon – ARK to LARK
anti-neutrino	condensed left-handed CW photon – anti-ARK to anti-LARK
electron	condensed anti-neutrino – LARK to SPARK
anti-electron (positron)	condensed neutrino – anti-LARK to anti-SPARK
proton	condensed positron – SPARK to QUARK
anti-proton	condensed electron – anti-SPARK to anti-QUARK
neutron	condensed positron followed by a condensed electron – SPARK plus Lark to QUARK
anti-neutron	condensed electron followed by condensed positron – anti-SPARK plus Lark to anti-QUARK

18 - Summary

1. Electromagnetic (EM) waves are loops.
2. Loop interaction is how forms are structured.
3. EM waves are composed of right-handed counter-clockwise rotating loops that are entangled with left-handed clockwise loops, forming a **wave-set**.
4. The various force fields are determined by the degree to which the electromagnetic wave/loops are condensed.
5. Loops and twists have the same energy value.
6. Fermion energy is loops times twists squared, same equation as $E = mc^2$. In FST the equation can be simplified to $E = O^3$ (whereby O = loops).

7. Considering a loop to be a line in a Sierpinski Triangle Fractal; a hierarchy of iterations of the triangles track the hierarchy of particles.

   To restate:

8. By counting the number of lines (loops) in each of the first seven iterations of the STF fractal determine the mass values of the first generation of particles.

9. The seven iterations of the Sierpinski Triangle Fractal have a progression of loop/mass values that are 3, 9, 27, 81, 243, 729, 2187. From these numbers all particles with masses between 0.511 MeV (electron) having 3 loops and the 1115.683 MeV (lambda hyperon) having 2187 loops, can be found as iterations of the Sierpinski Triangle Fractal.

10. Form and structure in nature is determined by the interaction of loops that form a three-dimensional spatial event called a fieldstructure that account for the fundamental properties of energy and matter.

   This is a highly condensed paper that focuses on one single problem in physics, which is: **How does nature account for the mass and energy values of particles?** What is the order that produces particles whose mass and energy have been found experimentally without the aid of a theory? This paper uses Field Structure Theory and Tometry analytics to resolve this problem of origins. Tometry introduces the idea of using interacting loops to account for structure directly applicable to the natural world of particles. Mass values can be based on structural models at the macroscale, iterated down fractally, modelling the organizing architecture of fundamental particles. As has been shown, the universe is composed of loops that unify all of nature into an evolving hierarchy of quantum structural platforms that are fields of force, each a derivation of the other. There is nothing separate from anything else. Reality is connected by loop architecture. Forms evolve as energy condenses, and devolve as energy is deployed. Action is made of right and left-handed rotations of electromagnetic wave/loops derived from the Plenum. Each form in nature is related to Plenum and exists because of how energy structures itself.

CHAPTER 2

LOOP ARCHITECTURE OF PARTICLES

19 - Loop Architecture of Particles

Tometry defines the circuitry of action based on loop and twists that are knotted into fields. Fieldstructures use that analytic to show how fields of action are stabilized to produce particle mass. Action is to Tometry what energy is to physics.

Energy is deployed loops.
Mass is condensed loops.

Field Structure Theory counts the number of loops involved in a form to get the mass number. The energy of that mass is the mass number squared. These relationships have been found empirically in fieldstructures. The mass squared refers to the number of times the loops rotate around its axis. A rotation of the loop counts as one full twist in the loop. Multiplying the loops and twists together is the total energy in the system. With the correct Tometry (geometry), the architecture of loops at the macroscale can be shown to be the same architecture of action loops used by nature at the microscale.

Structurally speaking, a loop in Tometry is a closed three-dimensional line of action. This paper will show how the material loops modelled at the macroscale having three-dimensionality produce the quantifiable forms and structures of particles. Fractally scaled down (*downerated*), fieldstructures predict the mass and energy values of natural particles in the microscale. Because Tometry reveals the natural way nature creates forms and structures at the macroscale, and because the universe is a fractal, iterating the macroscale model down fractally to model microscale structure is possible. Iterating up (*uperating*) to model the cosmic scale of structure is also possible. It is important to bear in mind as we go from the macroscale to the microscale to the cosmic scale, we are structurally following the fractal architecture of the universe. The scaling up or down of observations by physics revealed that reality became indeterminate. To solved that problem quantum mechanics was invented. What should have been concluded was that our notion about the structure of reality was improperly understood. Instead of replacing "certainty" with uncertainty, reality was telling us nature adheres to the "Multiple Certainty Principle".

The static models you will see in this paper have the same structural system that is found in the kinetic architecture of atoms and particles. It sounds fantastic, but it works! It works because the hierarchy of relationships of macroscale structures is the same hierarchy that exists between particles but at a smaller iteration of the fractal sequence. Loop analytics will show that it is possible to reproduce static macroscale structures, shown in this thesis, as kinetic macroscale structures. The proof of concept experiment awaits the participation of a willing plasma lab.

If a kinetic electron macroscale fieldstructure can be successfully produced, it means the way nature structures the universe at all scales will be empirically verifiable at the macroscale. The reason I feel confident fieldstructures are how nature structures particles at the microscale is in the fact that macroscale structures shown in this book accurately produce the numerical relationships experimental particle physics has determined. If the relationship of fieldstructures at the macroscale model, the relationships of the microscale, inversely, the microscale structures should be able to be reproduced electromagnetically at the macroscale.

20 - Assumptions

1. By knowing the structural hierarchy, the physical attributes of a mass and energy can be understood, modelled and predicted at all scales enabled by the fact the universe is a fractal organization.
2. The four known forces in nature are derived from the action matrix of the Plenum.
3. The Plenum is a universal chiral loop matrix and a fifth force.
4. Plenum is the only universal force. The other four forces are finite and limited.
5. The four forces of gravity, electromagnetism, strong and weak are field forces. These field forces are all condensed loop fields of varying magnitudes.
6. Plenum is a fully deployed symmetrical field not constrained by space and time. Space and time only arise with the formation of mass and energy.
7. The four condensed field forces are caused by the dis-equilibrium of chiral energy that develops when chiral wave/loops derived from the Plenum are asymmetrically condensed and/or deployed.
8. In structure physics, as in quantum physics, there is a unit of energy, a quantum. In structure physics that quantum unit is a loop.
9. A closed loop is a quantum unit.

10. Field forces are determined by the number of times the loop interacts with itself, or with other loops.

11. Interaction causes the loops to twist and twisting produces frequency.

12. When field forces interact, the Plenum is twisted and condenses.

13. There are two kinds of energy: (1) deployed centrifugal energy and (2) condensed centripetal energy.

14. Centrifugal energy is the kinetic release of condensed action/loops that allows the loops to untwist, expand, convert energy into spatial extension.

15. Centripetal energy is the condensation of deployed loops that lead to energy containment in a domain to produce mass, trading spatial extension for increased energy.

16. A loop having energy is thought of as a line of action (LOA) that has been rotated, either by bending the line of action into a loop or twisting the loop.

17. A line of action (LOA) that is twisted has topological attributes, a circumstance made clear if the (1-D) line is considered a plane (2-D) or a volume (3-D). FST considers a line to be three-dimensional.

18. A unit of boson energy is the twisting of the line of action a full 360° rotation.

19. A unit of fermion energy is the looping of the line of action. To make a loop requires a full 360° rotation.

20. Mass is the result from having the bending and twisting (torsion) of a line of action done in such a way that the torsion energy induced can be contained.

21. An interrupted loop (cut loop) reassembles back into a loop(s) through a process called in FST, *"The Heal-Cut Mechanism"*.

22. Loop energy cannot be destroyed. Through the Heal-Cut Mechanism, cut loops heal themselves and return to being a closed loop (torus) without a loss of frequency. This can be empirically demonstrated.

23. It is through knowing how energy can be knotted three-dimensionally that the forms and structures of nature can be understood and predicted.

24. It is through the interaction of a loop with itself, or other loops, that frequency forms and quantifies.

25. Loops behave as fractals. This fact enables a line of action (LOA) to be structurally modeled at the macroscale and then downerated to the microscale or uperated to the cosmic scale.

26. Though the forms change, the same structural principles at work in the macroscale operate at the microscale and cosmic scale.

27. Loops build hierarchy and produce the various platforms of structure found in nature.

28. Platforms of structure from smallest to largest complexities are all forms of loopage: Each platform has a minimum/maximum loop count.

    (a) Plenum (the action matrix, background-dependent substratum)

    (b) Electromagnetic wave (EMW)

    (c) Particle

    (d) Atom

    (e) Molecule

    (f) Cell

    (g) Etc.

Simplicity can produce complexity without negating continuity or cancelling provenance. All forms and structures are derived from interactions that occur to the universal **background-dependent** Plenum. Background-dependency means in Field Structure Theory (FST) that energy and matter are derived from the Plenum action matrix and are inseparable from that Plenum the way a knot is inseparable from the string on which it is tied.

21 - Tometry and Field Structure Theory (FST)

Tometry is the analytic that combines interacting loops to form three-dimensional knots that arrange themselves fractally to quantify form and structure at all scales. A three-dimensional knot is not a knot tied in 3-D space (a trefoil), but rather a knot that defines and creates three-dimensional space (a fieldstructure). Tometry uses macroscale structures to delineate microscale and cosmic scale form. Tometry is a field analytic.

Given the above assumption, the necessity is to show how the loops can produce form and structure quantitatively as well as qualitatively.

Field Structure Theory is the application of Tometry to understanding the natural physical world of form and structure, from microscale, to macroscale, to the cosmic scale – made possible by the fractal nature of form and structure. FST using Tometry's ability to model energy, provides a quantitative model for how boson waves form and then interact to create fermion particles, atoms, and molecular structures.

22 - Loops – The Energy-Centric View of Reality

The four axioms of Tometry are:
1. **Lines delineate action.** *Action is to Tometry what force is to physics.*
2. **Lines are loops.** *They have no beginnings, or endings, and cannot be destroyed.*
3. **Lines have dimension.** *They are neither infinitely thin nor are they without extension.*
4. **Lines interact.** *They do not intersect.*

The context is the Plenum. The Plenum is the action matrix from which all mass and energy events are derived and into which they decay. The Plenum is background-dependent. A loop interacting with itself, or other loops, produces a wave having frequency. To have energy, action must have frequency. To have frequency, a loop interacts with itself or other loops. In physics, energy is frequency times a unit of fundamental space/time, i.e., (E = fh), wherein "h" is Planck's constant and "f" is the number of times the loops rotate around each other.

A unit of action in Tometry has two forms.

(1). A loop constitutes a unit of action because to make a natural loop, the LOA (line of action) has to rotate 360° around its axis; see A in Fig. 2.39

(2). A full 360-degree twist of a LOA (line of action) loop around its axis is a unit of action; see B in Figure.2.39.

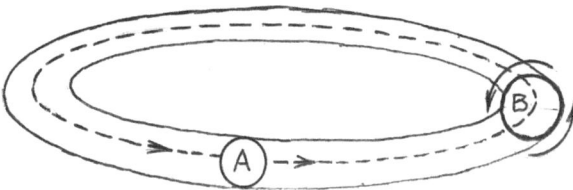

Fig. 2.39

In an analog clock with hands, the hour-hand moves one unit every time the minute-hand moves sixty units. The hour-hand is a loop and the minute-hand is the number of twists in an hour-hand loop. Both looping and twisting constitutes the total action in the system. The rest energy of a natural system is the number of loops times the number of twists squared. In physics the equation is E = mc².

As will be shown, the fact loops model action at the macro-scale does not limit its application to all other scales of action. The same principles of structure that operate at the macroscale also operate at all other scales.

FST shows how form and structure that can be modelled at the macroscale can be appropriately applied to any other scale when they are fractally iterated up (uperated) or iterated down (down-erated). Thus, it is possible to model the microscale world of particles, atoms and molecules with forms and structures modeled at the macroscale. The inevitable question is, **"How can a local material loop modeled at the macroscale, model microscale action?"**

It seems necessary to explain further why I believe a macroscale model can successfully define what is happening structurally in the microscale.

A loop made of materials in the macroscale can interact with itself to produces frequency by winding around itself or other loops. In the same way, a loop made of energy in the microscale can interact with itself to produce frequency. Material loops and energy loops utilize the same structural imperative. That imperative is inherent in the structure of the Plenum. The miracle of nature is that it has a way of keeping those interacting loops interacting perpetually; a proton and electron vibrate forever unless acted upon by an outside force. This fact alone requires an action matrix Plenum to exist.

Producing waves having frequency at the macroscale is the same in principle as the way nature creates frequency at all other scales. Particle loops are electromagnetic energy, which is the same energy that produces the strong and weak forces having differing degrees of loop/wave condensement. Macroscale field-structures are material loops that obey fermion statistics of interaction and Pauli's Exclusion Principle. Macroscale models made with materials are the fermion equivalent to the boson energy structures found in particle physics.

Boson waves, making up particles, are kinetic in nature whereas fermion waves modeled in the macroscale are static.

Static means in FST the wave is not moving because the wave is everywhere the loop exists. Instead of having one particle circuiting in a loop, at the macroscale where the structure is modeled with materials, there are billions if not trillions of particles in the loop making the loop solid. The structure is the same; the number of particles in the structure differ.

In a kinetic wave at the microscale the impulse is moving in a circuit, all be it at near the speed of light, in the case of an orbiting electron, whereas in a static macroscale wave, such as the structures seen in this book, the impulse is everywhere on the wave so it doesn't appear to be in motion. What can be seen happening with a fermion wave (made of rope for instance) mirrors what happens in boson waves which cannot be seen. Furthermore, because the separate wave forms of a fermion wave are superimposed on each other, they both occupy the same domain. Boson waves can entangle without causing interference (see Fig. 2.40). When the waves are looped, they cannot be disentangled, but they can be separated. Both have the same circuitry. The boson waves interacting have the shared axis at the center of the wave's helical rotation. Fermion wave/loops have their shared axis on their outside surface of each wave and also remain entangled. Both fermion and boson describe a way in which a collection of interacting entangled wave/loops may occupy the same parallel energy states by sharing the same axis of rotation. This is also why in the 1st shell of an atom the boson electron wave can be rotating CW or CCW, even though the fermion electron particle rotates CCW.

Energy loops interacting three-dimensionally can knot together and form fermion structures, be they micro or macro in scale. When the wave length of electromagnetic loops matches the diameter of a particle such as a proton at 10^{-16} m, the wave condenses and knots into a particle. Why it does this at that particular distance has something to do with the inherent structure of the Plenum; the exact nature of which needs to be further understood.

The phenomenon of radiant waves collapsing into a particle is at the heart of how mass forms from energy. A hypothesis of FST is that the same conversion of energy into mass that occurs at the microscale of particles, can occur at the larger macroscale. Even though using material fermion loops, with FST it is possible to gain insight into the behavior of boson loops of energy. **Material fermion loops at the macroscale can reveal the circuitry of boson energy at the microscale.** The proof of this assertion is whether or not modeling energy at the macroscale produces the same ratios of mass and energy found in microscale particles. If the macroscale model can do this, then the assertion has strong support for its veracity.

Fermion mass is contained boson energy. Mass is localized condensed EM energy. When speaking about energy, I am referring to non-local boson EM energy. When speaking about mass, I am referring to localized boson energy, which is condensed energy. Condensed boson energy behaves as fermion mass. All energy is in the form of torsion. The energy created by twisting a string is stabilized by making the twisted string into a loop without allowing the twisting to untwist. If the string has endings (i.e. is not looped), it cannot hold a twist and though it may be twisted, it will untwist. By twisting a string into a loop, the twist energy is locked in. Each time the line of action/loop is twisted 360°, a unit of energy (a quantum unit) is added to the line of action. **Energy in simple terms is rotation.**

Shown in Fig. 2.40 is how FST understands a loop (only a section of the loop is shown). A loop entangled with itself or with another loop is called a wave-set (or loop-set).[4] Wave-sets automatically have clockwise and counter-clockwise rotation occurring at the same time. A wave-set has chirality, but does not exhibit chirality when the wave-set is deployed. Chirality only becomes

---

4   *The term "wave-set" refers to the form when speaking in the context of physics. The term "loop-set" refers to the same form when speaking in the context of Tometry.*

apparent when one side of the wave-set condenses. In nature, the electromagnetic wave is a wave-set composed of interacting clockwise (positron wave) and counter-clockwise (electron-wave) rotating loops. In Figure 2.40, the red wave is moving right to left and the black wave is moving left-to-right in a 3-D spiral that wraps around each other.

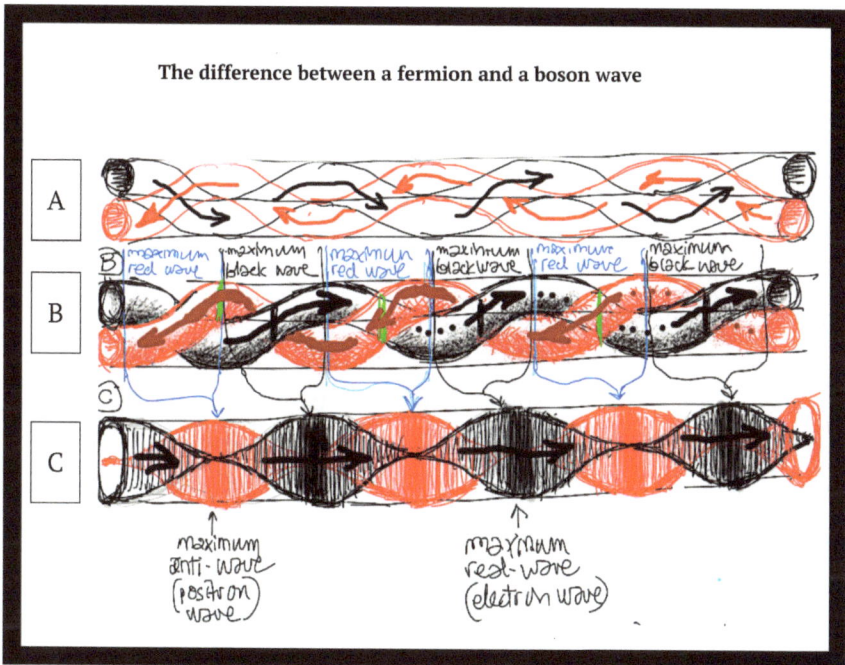

The difference between a fermion and a boson wave

Fig. 2.40 – Boson waves are wave-sets that can have the two chiral waves occupying the same domain, i.e., the same internal axis. This illustration assumes that the boson wave, here called an electron-wave drawn in black is a "real-wave", meaning we can empirically observe it. The wave drawn in red is an "anti-wave"(positron-wave). It is not discernible by any real-wave means. Both waves move in the same volume going through each other in opposite directions. The "real-wave" moves forward in time and the "anti-wave" moves backward in time.

Waves wave because the anti-wave (the positron-wave in red), which we cannot see, opaque's the real-wave (in black), which we can see. Because anti-waves are entangled with real-waves, the anti-wave, and the anti-matter produced by anti-waves, cannot be seen by a real-wave sensory system, such as the human eye or real-matter machinery. Because the waves are wrapping around each other, the wave in the foreground opaque's the wave in the background in a repetitive manner preventing both waves from being fully visible at all times. Instead of seeing waves as continuous with a fixed amplitude, what is seen is a pulsating wave; a wave that grows and then diminishes in amplitude repeatedly. This means the red anti-wave in Figure 2.40 is alternatively masking the real-wave, which makes the real-wave appear to be pulsating and because of this pulsation, the wave has frequency. The process is known as *cloaking*.

In item (A) in Fig. 2.40, the red and black wave/loops move through each other without interference. Though depicted as in separate domains, they are not separate in the boson state of being pure energy. In item (B) in Fig. 2.41, the two waves are rendered as solid forms wherein each wave has its own volume and opacity. This is done to show there are two different waves, but as noted in (A) because they are boson waves, they share the same domain.

Waves of energy (boson waves) don't look or operate separately as Figure (B) would suggest. The rendering in (C) shows the net-effect of chiral boson energy waves interacting. The red boson energy wave, called the anti-wave, opaque's the real-wave in black so that the real-matter viewer, who only experiences the black wave, sees the black wave as pulsating.

Boson waves are electromagnetic (EM) energy. Mass is the bound state of electromagnetic energy (EME)

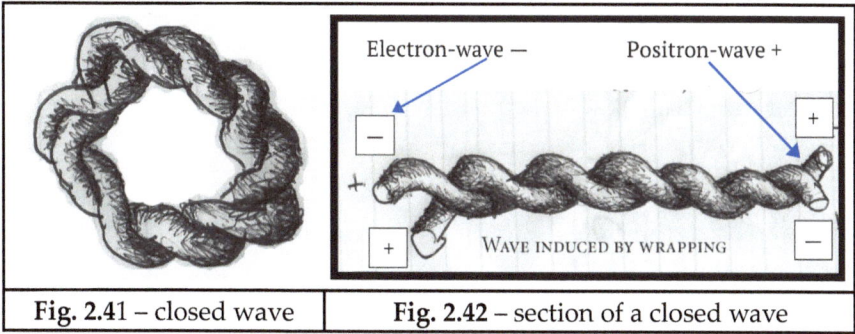

Electron-wave —   Positron-wave +

WAVE INDUCED BY WRAPPING

| Fig. 2.41 – closed wave | Fig. 2.42 – section of a closed wave |

Fermion waves are entangled wave-sets that have a chiral relation-ship. A fermion wave has the axis of the form on the surface of the loop material or at a median distance from the two related loops.

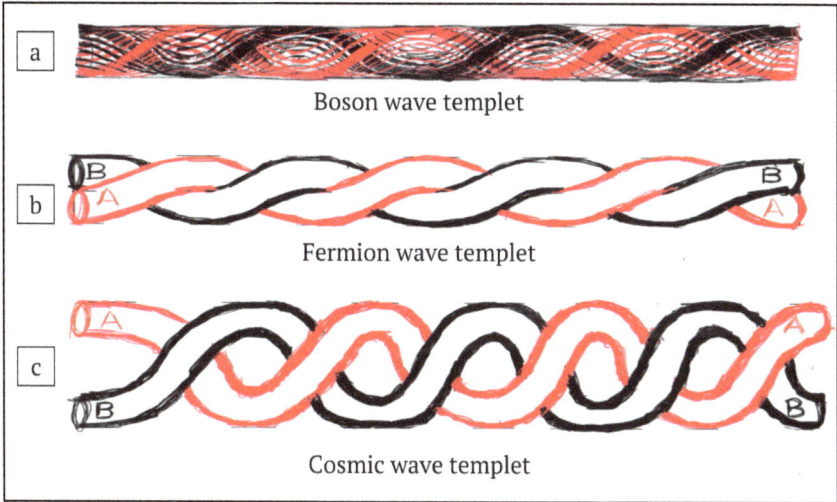

a   Boson wave templet

b   Fermion wave templet

c   Cosmic wave templet

Fig. 2.43 - The three forms of wave entanglement

(a) Waves share the same axis at their volume centers...the microscale model.
(b) Waves share the same axis at their volume's surface...the macroscale model.
(c) Waves share the same axis at a distance beyond their surface volume...the cosmic scale model.

Our sensorium only recognizes matter and how energy affects matter. We can't see energy in any other way. A fermion wave at the macroscale shares the same axis as a boson wave, but the axis is on the surface of the wave form, not at its center. The axis of macroscale fermion loop is where the loops touch, or alternatively, they are a mean distance to each other. For a wave-set to be entangled, each wave must share the same axis, which is the point of contact between the two loops of a fermion wave, or the central axis of the two loops of a boson wave. The two loops of a fermion wave-set are entangled, but occupy different spatial domains within the wave-set. The Pauli's Exclusion Principle applies. When modelling with materials, requires the waves to separate while being entangled and yet share the same axis (see Fig 2.43-b). Waves made with materials are fermion waves. Waves made with energy, such as electron rings, are boson waves. Nature does not separate the two loops of the boson wave. It is only when the loops condense do the loops develop separate though connected domain as can be seen in a 3-D fieldstructures.

In summary, there are two kinds of EM waves, (1) Fermion Waves accounting for how materials have wave characteristics and are localized forms of energy, and (2) Boson Waves that account for how waves can pass through each other and are non-local forms of energy.

Field Structure physics is the physics of looping and twisting.

A "whirly-gig" exemplifies the chiral motion mechanism; by pulling the strings in opposite directions, the whirly-gig spins in position. Energy fields have these chiral forces pulling in opposite directions and accounts for particle spin. Mass can be thought of as a local event being pulled equally in opposite directions.

Particles spin because chiral waves associated with a particle, pull it in opposite directions as the whirl-gig demonstrates. Energy manifests as the release of torsion that has been twisted into a loop. Releasing energy relieves the stress that has been induced by twisting the line of action. By making a line of action into a loop, the energy induced into a wave by twisting is stabilized. By interacting loops properly – so that they make a fieldstructure – the energy forms a knot and achieves stability. Two things can happen to a loop. It can be bent into additional loops.

Fig. 2.44	Fig. 2.45	Fig. 2.46

Fig. 2.44, 2.45, 2.46 – Looping One loop can have many loops. The loops can be inside or outside loops.

OR ... the loop can be twisted around itself.

In these rope examples, a single loop interacts by winding around itself to produce a wave having an odd-number frequency; all wave forms made with a single loop have an odd number frequency. Each node is a full 360-degree twist of the line of action. Fig. 2.47 shows the twistings of the loop deployed so that the twisting is spread around the loop/wave evenly. In Fig. 2.48, the loop/wave has condensed its wave nodes, gathered them together as tightly as possible. Fig. 2.49 is a detail of the twisting revealing the hidden chirality of the wave that was not obvious when both sides of the wave-set were deployed as in Fig. 2.47. Note that the single circular loop naturally wants to form a linear loop structure when condensed. I did not impose the linearity of the wave other than compressing the wave as much as possible.

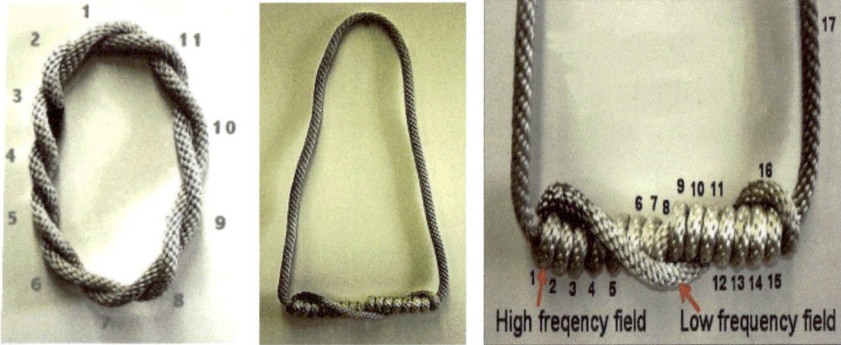

Fig. 2.47 – deployed	Fig. 2. 48 – condensed	Fig. 2.49 – detail of Fig. 2.48

The structural hierarchy of forms having localized mass begins with the loop. The structural hierarchy of non-local energy begins with the twisting of a loop. The way to induce twisting is by two methods – making a loop or by twisting the loop about itself. Both looping and twisting involve 360-degree rotations and each rotation constitutes a quantum of energy.

Two loops interacting are called a wave-set.

In Fig. 2.50 and 2.51, two chiral loops interacting with each other produce a wave-set having an even number frequency. All two-wave interactions have an even number of nodes. Fig. 2.50 is a deployed wave-set made with two loops. Not discernible in Fig. 2.50 is the fact that one loop is rotating CW and the other CCW. In Fig. 2.51 a wave-set has had one side of the wave-set condensed and in so doing, carries all the frequency in the wave-set with it, except for one unit of twist that remains in the deployed wave. This is what happens in physics when a non-local wave-set collapses to form a particle having both a locality and an extended non-local charge field. The particle part of the form is where all the energy in the wave-set has condensed to a particular place

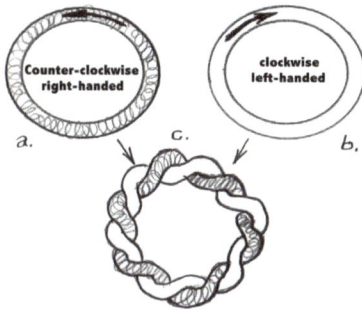

Fig. 2.50 – A wave-set deployed

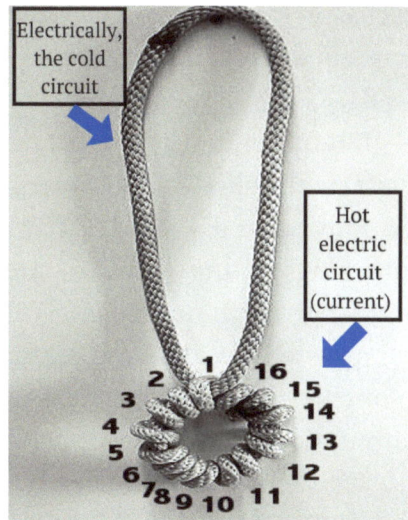

Fig. 2.51 – a wave-set condensed

and becomes a local particle event, leaving an extended non-local field of opposite handedness surrounding the particle field. How waves structurally collapse, as physics calls the process, has gone unnoticed probably because no one in physics who has the microphone has played around with ropes, believing there is no connection to what goes on between macroscale ropes and microscale lines of action. FST physics will reveal that what exists structurally at the microscale with kinetic energy (EME) can also happen at the macroscale with loops of static ropes, *and* by implication suggest that macroscale electron loop structures are possible. With a successful proof of concept experiment, the physics will know how microscale particles can be recreated at the macroscale.

Fig. 2.51 is a template for an electric motor. The deployed circuit is energized by the magnetic field surrounding the electric field. The magnetic field squeezes the electric field and thereby induces pressure on the electric field. A charged particle in the electric field will be accelerated in proportion to the number of twists in the magnetic wave/loop.

It is by condensing (what in physics is called a *wave collapse*) that the chirality of the wave becomes apparent. It is from the condensing of a wave-set that chirality and charge arise. When both sides of the wave-set are deployed, the electromagnetic wave (light) has no charge. Only when one side of the wave-set collapses will charge and handedness appear. Since a particle is a condensed wave, a particle will have a charge field associated with it that will be the opposite handedness as the charge field.

The neutron does not have a deployed charge field as does the proton. Because of that, it has no charge. In the neutron, an electron has condensed to nuclear dimensions and become a gluon. The gluon's job is to connect protons. There cannot be two protons in a nucleus, whereas there can be one proton and one neutron. A deployed neutron is unstable because the condensed electron has too much energy to be constrained to the neutron's proton core. In other words, there has to be two electrons linked to two protons, but one of the electrons will condense into the nucleus and become a gluon that the two protons can share. When the gluon is close to one of the protons, that proton is a neutron. A proton becomes a neutron as the gluon (condensed electron) alternates between the protons.

The collapse of an electron into the nucleus is, in a sense, a charge field collapsing into a particle field. This produces a neutral charged particle. Since the two fields are now together, they are both condensed. To have a charge field, the electron would have to deploy, while the field object particle (proton) to which it is attached, remains condensed. Should the neutron find itself outside the nucleus, the condensed electron (the gluon) that made the proton into a neutron has too much energy and flies off as a beta ray. A neutron without its electron/gluon becomes a proton, and the gluon is now a high-energy beta ray. The high-energy electron, now a beta ray electron, has too much energy to be held by the proton magnetically, preventing the formation of hydrogen. If the proton could hold the departing electron, hydrogen would form, but alas, the electron cum beta ray is too powerfully kinetic and flies off. Neutrons cannot form hydrogen when it decays even though the decay products are a proton, electron and neutrino.

In the case of the electron particle, should it condense the way a positron condensed to become a proton, it will become an anti-proton with a negative charge, and will be 1836 times more energetic when it was deployed as an electron. When a wave condenses (collapses), the deployed wave and the condensed wave will have opposite charges. The collapsed wave will be spinning either clockwise or counter-clockwise and be either real-matter or anti-matter. In real-matter, the condensed wave has a positive charge because it is the positron-wave side of the wave-set that is condensing. That leaves a negatively charged electron-wave side deployed.

Conjoined EM loops make a quantum system. While they cannot be separated, there can be a wide separation between that portion of the wave that is condensed and that portion of the wave that is deployed. The separation can be as big as the universe. The Aspect Experiment done first by the French physicist, Dr. Alain Aspect, can be understood as resulting from the way loops are entangled even though their condensed forms can be widely separated spatially and intimately connected temporally at the same time. In the Aspect experiment, if you know what the handedness of energy is at any distance from its source, you know instantly the handedness of its quantum partner. This entanglement is easily seen in rope models.

Waves that are not entangled in a wave-set are not a part of the same quantum system. As mentioned, radiant EM waves have neither charge nor are they affected by a charge field they may pass through. Something that has *charge* means in FST that one side of the wave-set has been condensed. If there is no restraining mechanism, a condensed wave/loop will deploy back over the deployed wave to distribute its wave frequency equally between the two waves of the wave-set. This ever-present effort to deploy and equalize the energy is what entropy is all about. It is nature's demand that there be symmetry and balance.

A neutral EM loop-set such as the neutron, has no charge, because both the negative electron-wave and the positive proton-wave are condensed. The reason the electron-wave, when it condenses to form the neutron, does not have the same energy as the positron-wave is, as previously mentioned, because the positron wave in proceeding the electron wave to condense, takes all but one unit of the loop-set's loops with it to form the proton. Thus, when the deployed electron-wave condenses to the proton to make a neutron, it adds only .06% of the total energy in the proton, but that is enough to convert the proton into a neutron. That is why the neutron is only one electron and one neutrino mass heavier than the proton (938 Mev for the proton versus 939 MeV for the neutron). The condensing electron also brings its tiny electron-neutrino with it. Adding the mass of an electron and a neutrino to the mass of a proton equals the mass of a neutron.

23 - Fieldstructures

Mass is a count of loops
Energy is a count of twists
Both are 360° rotations of the line of action

The definition of fermion mass in FST is a form that has a minimum of three loops that are knotted together and define a 3-D spatial form. For example, mesons are composed of two quarks, one is a real-matter quark and the other is an anti-matter quark. This is an unstable combination). The reason hadrons are thought to be made of quark loops that do not exist as condensed loops outside the nucleus is because the three quark loops that are knotted together and held together by gluons (condensed electron-waves) become bosonic radiant gamma electromagnetic energy. Decaying quarks do not become particles. They become radiant gamma radiation and hence are non-local. Quarks decay into their radiant EME forms of energy when the nucleus receives more energy than the strong force can manage. The difference between a hadron and lepton is in the degree to which the EM loops are condensed. Leptons can condense to quarks and become hadrons, but quarks can't deploy to become leptons. The deployed energy of a quark is too great to become a lepton. Deployment of a wave and condensing of a wave are not therefore symmetrical. Their arrow of time goes only forward, not backward.

Figure 2.52 shows how loops create interactions that have varying degrees to which they are condensed. Quark loops are maximally condensed EM waves and thus are high-frequency gamma waves. Each additional figure below shows figuratively degrees of condensement and their names.

Ark	Quark	Spark	Lark	Bark	Aum
Plenum –	Gamma ray	X – ray	+ EM waves	–EM waves	Plenum +
$(1.6 \times 10^{-35}m)$ Minimum event	$(10^{-16}m$ to $10^{-12}m)$	$(10^{-12}m$ to $10^{-8})$	$(10^{-8}m$ to $10 m)$	$(10$ to $10^8 m)$ Maximum length unknown	the cosmic wave Maximum event
Planck particle	Nucleons	Electrons	Neutrinos	Photons	Loopsum

Fig. 2.52 – **Loop Families** – Take this illustration above as a schematic generalization. The loops show how the amount of condensement and/or deployment of loops produce a unique area of interaction that is the central circle created by the overlapping loops. This central area indicates how condensed is the field energy. The volume of a particle created by interacting quarks is 10^{-16} meters. The volume of the Aum loop is the size of the universe which is thought in FST to be 10^{+64} meters. However, almost certainly there are realities smaller and larger than these numbers. **The implication is that all energy forms in the universe are degrees of loop condensation.** Light is a loop and everything made can be made with light, i.e., EME. This graphic shows how loops relate to particles. Particles are simply condensed light (EME). To accurately define a human being, along with the rest of creation, would be to understand the whole shebang as condensed electromagnetic energy.

In Figure 2.52, the time arrow goes from left to right. This is the direction in time that leads to entropy. The right to left time sequence is syntropy. Entropy is how the world looks from the physical perspective. Syntropy is how the world looks from the metaphysical perspective. The time arrow here illustrated helps explain why by not including metaphysical time, the so called "real time" considered by physics is only half the story of time.

Condensed Aum loops are Bark loops
Condensed Bark loops are Lark loops

Condensed Lark loops are Spark loops
Condensed Spark loops are Quark loops

Loops condense non-local energy to localized forms (particles) in a step by step syntropic (large loops to small) progression, but when these loops deploy, they deploy entropically **directly** back into the Aum loop. Building matter is climbing steps up from the ground. Destroyed matter leaps back directly to the Plenum. The reason the time arrow doesn't go backward is that when a particle loop is completely destroyed as in the case of a real-matter/anti-matter interaction, the particle doesn't travel the step by step progression it took to form. There is no arrow of time to follow back over the step by step hierarchy taken in the formation of matter. The progression forward in time is step by step going from A to B to C and so on to Z. The progress backward in time is instantaneous; a jump from Z to A with no stops in between. Thus, quark loops (hadrons) can't become electron spark loops should a nucleon decay. They have too much energy. Condensed loops deploy from a particle state to radiation instantly, such as in the real-matter/anti-matter annihilation. Loops condense into particles from higher densities to lower densities through a process of decay, such as the lambda decay sequence.

In standard physics, first generation particles, are made of three quarks that are the maximum condensation of the EM loops. I say "maximum" guardedly because there are condensations of energy loops more energetic and smaller in volume than the proton. Such particles are being searched for in Europe by the LHC at Cern, France.

24 - Space

Space is produced by three-dimensionally knotting loops. The forms that create space are fieldstructures. The notion that mass fields define space may be counter-intuitive. Observationally, we are at the boundary between inside space and outside space. We can see the fieldstructure form in particles and atoms, but it is not so easy to see cosmic size fieldstructures. FST forced me to see that we are inside a fieldstructure when looking out at space. Stars and galaxies are the vortices (*vertices* in Euclidian talk) of a fieldstructure, where action interacts three-dimensionally. It is perhaps the enormity of the stars in a galaxy that disguises the fieldstructure form. When the circuitry of the energy in the system is simple as is the case of the Structor or SuperStructor, seeing how the vortices define space is easy. Seeing a star system or galaxy as a fieldstructure requires seeing that the large amounts of mass involved, creating strong gravity, flattens the energy sphere into a disk shape. This flattening of the fieldstructure is maybe gravity's primary contribution to the shape of a galaxy. In particle fieldstructure, gravity is too weak to flatten the spherical orbital motions of a particle. The rotational motion of the stars in the galactic plane is determined by EM forces that are far stronger than gravity force. The rotation of a body is determined by EM centrifugal forces, not gravity. Gravity accounts for the centripetal forces.

Space is not neutral, nor void, nor without attributes. From the Plenum, energy is made. From energy, matter is made. Energy and matter make space. Without the Plenum interacting with itself, there is no energy available to make matter. Fieldstructures are condensed Plenum that produces space and time by producing mass and energy. Mass defines space, while at the same time energy is marking duration, i.e. time. Microscale fieldstructures are stable in and of themselves. They behave as does an instant. The action field supplies the connectivity and the stability. Each of the four fields are condensation of loop energy derived from the Plenum. The purpose of a field is to maintain equilibrium and at the same time maintain a balance between mass and energy as implied by formula $E = mc^2$ (m is the condensed loop field and c^2 is the deployed energy field that carries the time component). Macroscale fieldstructures can also be stable by relying on the material connectivity of the loop and by using the same structural architecture used by microscale forms.

The loop is the form taken by a field. All form is field generated. It behooves us to know that a field can be as simple as a tetrahedron field or as complex as a human field. At the base of it all is the mighty loop. In all this expanse of forms in the universe, it comes down to the number of loops and the frequency of those loops. Loops naturally seek system-wide equilibrium. The reason nature seems to never obtain universal equilibrium is the inadvertent omission of the metaphysical behind the physical. Metaphysics pervades the physics. Metaphysics brings equilibrium by the integration of the anti-world and the real-world. This is not a problem for the structuralist who demands the sum total of any action/reaction to be zero. Real matter divided by anti-matter equals equilibrium equals zero energy. It is tantamount to being the solution to the question, "Why". Physics has not been wrong or at fault, rather it has been simply incomplete.

The dual of physics is metaphysics. Metaphysics is needed to balance the cosmic books. Balancing the cosmic books is the reason we do physics. The income side of the books is metaphysics and the expense side are physics. When the knot gets around to discovering it is string and the string realizes it is tied into a knot, the books will be balanced. To understand the Plenum, physics and metaphysics have to work together. Matter and anti-matter are realities; the only reality that can give us a noble and complete cosmology. Fieldstructures provide the model.

25 - Energy as Loop Circuits

The fieldstructures pictured in this paper are static fermion circuits using materials to carry torsion energy instead of using EM currents. The fact fermion circuits are structurally equivalent to boson circuits of energy is a revelation. This shows that structure when understood is a stand-alone science applicable to anything having structure, be it material or immaterial.

Fieldstructures define the action circuits of matter. These circuits are deterministic and precise, not a vague generalized amorphous probability cloud type approximation as quantum mechanics describes. If you could cool down the circuitry of an atom having a certain mass/energy, slow down the action in the form, eliminate the processional motion of the circuits to rotate around the form, and prevent incident inputs of EM energy from impinging on the system, the circuitry of the action would demonstrate the fieldstructure form used by nature.

Since directly observing an individual cold atom is not technically possible, there is another way; to approach this problem; reproduce the atom at a larger scale. Scaling a form up slows the form down. Scaling an atom up (uperating) twelve powers slows it to a virtual stop. FST maintains that in building a fieldstructure, the structural system that is moving close to the speed of light, such as does particle waves of an atom, is slowed to a halt in the macroscale fieldstructure. Admittedly this seem preposterous. It is a possibility that seems worthy of exploration.

Increasing mass condenses the energy.

Increasing the energy deploys the mass.

The force in macro-structures shown in this book is due to the bending, compressing and stressing of material loops each striving to expand in order to relieve its torsional stress. However, because the form is a 3-D knot, the loops are blocked from doing so by the other loops who are at the same time trying to unbend in opposing directions. We have a three-dimensional world because two loops in separate axes are sufficient to electromagnetically block a third loop from returning to a stress-free circular loop. Our world is stable because the function of a field is to establish equilibrium within a stressed environment.

Structurally, the particle/wave paradox is due to the faulty concept we have about "what is a particle?". It is in defining a particle that opinions are numerous. Untold volumes have been written discussing the difference between particles and waves. **The fieldstructure view is that a particle is a field of interacting loops knotted in three-dimensions and defines a spatial volume.**

Fig. 2.53 – The morph from particle to wave is a matter of a loop condensing or deploying.

By knowing the correct structure, then the problem of how a wave can become a particle becomes obvious. To add to FST's definition of a particle, it can be said a particle is where three wave-sets (wave-sets are also called loop-sets) with gamma frequency interact in a holographic manner. Particles exist when loop interactions are being viewed by a larger loop system. We see atoms and particles as local fields having mass (Fig. 2.53a) and we see non-local wave structure (Fig. 2.53b) as fields. If we were a particle, we would view molecules as nearby galaxies.

Where EM loops interact, standing waves form. Standing waves that achieve three-dimensionality are particles and behave as hologram type structures. At gamma wave intensity where the distance between nodes of the three interacting loops match the diameter of a nucleon, a particle forms. The gamma wave has the frequency length that is the same as a proton's diameter. When the wave frequency of three loops matches the particle's frequency, the waves condense into a particle. They condense because of the electromagnetic dynamics that exist between the left-handed and right-handed loops. The left-handed clockwise loops condense if they are cloaked by the right-handed counter-clockwise loops. The left-handed loops with their positive charge condense to the nucleus and right-handed loops are deployed away from the nucleus. This produces an atom with a positive charge nucleus and a negatively charged field surrounding it.

At the microscale of particles, the deployed action field of a particle has a spatial volume 100,000 times greater than the volume of the nucleus. The energy is correspondingly 1836 times smaller when the loops have expanded 100,000 times, in volume. Models at the macroscale, such as shown here, have those proportions greatly compressed 100,000 times so as to appear as they do in the static fieldstructure model in Fig. 2.54. Building structure at the macroscale and then down-sizing through iterations to the microscale, suggests that the universe is a structural continuum. Time and space coordinate with energy and mass so everything is in balance at all scales of manifestation.

Looking at the world of micro-physics, an enlargement of the spatial energy field, requires a deployment of energy, which is to say the frequency of the field decreases as the volume of its spatial extension increases. Frequency in FST is the number of nodes in a loop/wave. The loop/wave can increase volume, while the frequency (number of nodes per unit of length) will decrease. The number of nodes will remain constant, even though the frequency decreases. The complexity of the nuclear polyhedron determines the number of the interacting loops and the number of times those loop rotate around their axes.

Frequency in standard physics can have different meanings. The frequency number can mean the number of waves passing a receptor in a period of time. This is not measuring anything about an individual wave. Frequency of a wave-train is measuring the number of waves (wave-fronts) passing a measuring device in a period of time. Physics treats wave-trains as being linear. FST is looking at a single spherical wave, not a wave train.

Boson waves are said by some to lose energy when they travel. That is because the wave is expanding and the nodes on the wave are being attenuated (stretched out). The wave is not actually losing nodes, that is, it is not losing energy *in toto*. The same number of nodes are in the wave when it began expanding as after it had expanded a light year, but the distance between the nodes increases as the wave's diameter increases. In a sense, the tired-light enthusiasts and the "light-doesn't-lose-energy-in-transmission" enthusiasts are both right. During transmission the deploying wave would seem to lose energy because the frequency is lowering as the wave spreads.

Fermion waves do not loose energy as they travel, because they are packets of energy contained within a local domain by which the containment mechanism (knotting) prevents the wave from expanding spatially. Fermion waves are *place particles,* meaning they exist in a place; their energy is not thinned-out by motion for the simple reason fermion waves don't expand. Boson waves are not knot-bound to a domain and are free to expand spatially. The relationship between a deployed radiant boson wave and a condensed particle fermion wave is that one can morph into the other when the circumstances permit. While we don't normally see the mechanism of proton morphing into gamma waves, macroscale fieldstructures can look at this process without being constrained by space or time. These ideas about waves and particles are not found in the Standard Model of physics.

The distance between nodes determines the wave angle. The proton and positron can have the same number of nodes in their deployed wave/loops form, but when either is condensed, the loops have vastly different spatial extension and yet collectively have the same number of nodes in total. For instance, when the positron wave of the electromagnetic wave-set condenses, the nodes on the loop that determine frequency are squeezed closer together to produce quarks with a frequency 1,836 times more energetic than the waves are in the lepton configuration. In terms of MeV, the proton is 3,370,896 (18362) times more energetic than the positron particle.

All waves are closed loops. Thus, it is correct to say that a quark wave is a highly condensed radio wave. The difference between an EM energy wave (boson wave) and the fermion waves of a mass depends on how condensed the waves have become. If this is true, then a radio wave could have once been a gamma wave. This implies the observable universe is a closed system, and that light (EME) changes frequency as it propagates.

26 - The *Structor* – The Fieldstructure Form Taken by Particles

The fieldstructure family of form that defines particles is the Structor platform. Structors are the simplest of all fieldstructures. In Figure 2.54, three loops interact to make a three-dimensional knot. This fieldstructure knot is a new geometric family of form found in the skew geometry of Tometry.

Loops knotted together three-dimensionally define a deployed charge field and a condensed polyhedron nucleus of the opposite charge. In Figure 2.54 the nucleated polyhedron is a tetrahedron. Any polyhedra can be circuited with loops. Fieldstructures have circuits of energy (loops) that can spin clockwise or counter–clockwise. In Figure 2.54 the fieldstructure is shown spinning counter–clockwise.

In terms of forms that can be seen empirically, what is seen in atoms, molecules and all empirical forms, are counter-clockwise (CCW) spinning electrons and clockwise (CW) spinning protons, while the neutron has an action field that spins both ways and hence has no charge.

While Figure 2.54 shows the visible empirical particle form, Figure 2.55 & 2.56 show what is actually going on in nature. In Figures 2.55 & 2.56 these forms are called *Ambi-Structors*, both the CW and CCW waves are present. In Figures 2.55 & 2.56, the CW spinning circuits (white tubes) are over the top of the CCW spinning (blue tube) circuits. That means the Structor is an anti-particle, because it is the CW positron wave that is on top of the CCW electron wave. Were the CCW circuits covering the CW circuits, the particle would be real-matter. In observing real-matter loops, we are only seeing one side of the wave-set. The other wave-set side is cloaked by the wave that is outward showing.

Real-loops and anti-loops are present in all forms. This assertion is not found in the Standard Model of physics. The real-matter side of the wave-set is **cloaking** the anti-matter side. Hence, empirically we see Figure 2.54 and not Figures 2.55 & 2.56. This is a drastic revision of particle structure. What is gained is an answer to the haunting question that has confounded physics for a hundred years: *"What happened to anti-matter?"* The answer is anti-matter is hiding behind real-matter.

In an atom, the CCW waves and the CW waves have separated so that we see them both in the model, but in different places with different degrees of condensement, even though they still are in the same wave-set. Both real-matter and anti-matter exist concurrently and normally remain inseparably attached to each other. When the positron wave and the electron wave are both deployed we only see the electron wave. When the positron wave condenses and the electron wave is left deployed, then we see the condensed positron wave as a real-matter particle. The anti-wave (positron wave) becomes visible because in condensing it is no longer cloaked by the electron wave. We can see an anti-particle when it condenses but not when it is deployed.[5]

The three loops of a fieldstructure seen in Figures 2.54 are the three quarks of the Standard Model responsible for neutrinos, leptons, and hadrons (or as FST calls them respectively, *quarks, sparks, and larks.*) The Standard Model does not recognize leptons, and neutrinos as having internal structure, nor are they understood as loops. Field Structure Theory (FST) does. In FST, any form having mass has loops and EME (electromagnetic energy) are loops.

Figure 2.54 is the same structure as Figures 2.55 & 2.56, but is without the anti-loops (anti-waves of anti-matter). Figures 2.55 & 2.56 have two deployed boson loops residing in the same orbit; the **electron loop/wave** and the **positron loop/wave.**

---

5   Isn't nature marvelous?

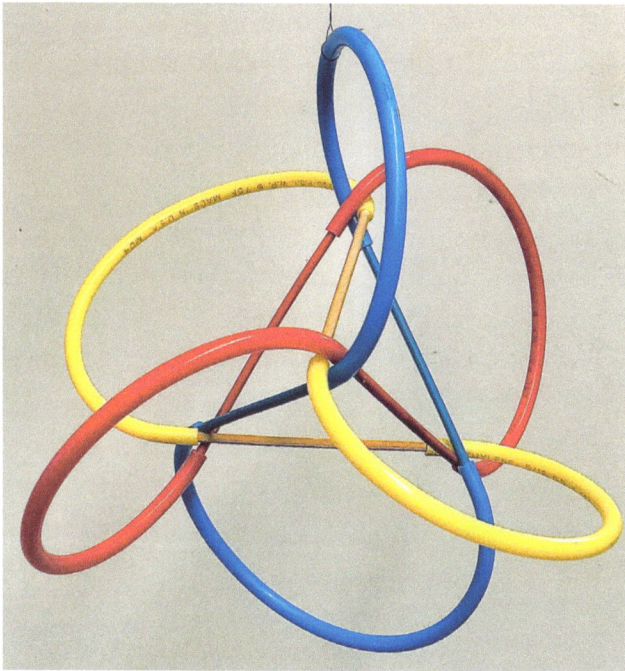

The family name of the form is .... . "Fieldstructure".
The genus name Fig. 2.54 is.......... "Structor".
The specie name of Fig. 2.54 is ...... "Tetrahedron".

Fig. 2.54 – The fieldstructure action field produced by three individual loops. This particular form is a fermion matter/wave called a **"Tet-Structor"**. The form has a CW spin which makes it an anti-particle. A CCW spin would make it a real-matter particle. The CW spinning Anti-Structor is the mirror image of the CCW spinning Real-Structor. The Structor holds its configuration by having each loop blocked from relieving the stress of being bent by two other loops trying to do the same thing in an opposing direction. No glue or fasters are used to hold this structure together. It is a topological localized knot of energy.

The examples shown in Figures 2.55 & 2.56 have three loops rotating counter-clockwise, and three loops rotating clockwise producing a tetrahedronal fieldstructure nucleus. Anti-waves do not interfere with real-waves in the same quantum system. In the laboratory physics creates a situation where a real-matter (RM) particle is separated in the quantum sense from an anti-matter (AM) particle. The AM flies off at high speed and collides with another real-matter particle. Because the two particles are in separate quantum systems, they annihilate and become CW and CCW gamma EM energy. The conclusion physics reached was AM and RM cannot live together, when in actuality they can live together when in the same quantum system. If you create from EM gamma energy a RM/AM particles at or near cold temperature, the two particles will not fly apart, but will magnetically conjoin and form the fieldstructure in Fig. 2.55 & 2.56.

Of primary importance about fieldstructure form is that in nature the three boson waves/loops, which are the three quark loops, create a fermion nucleus.

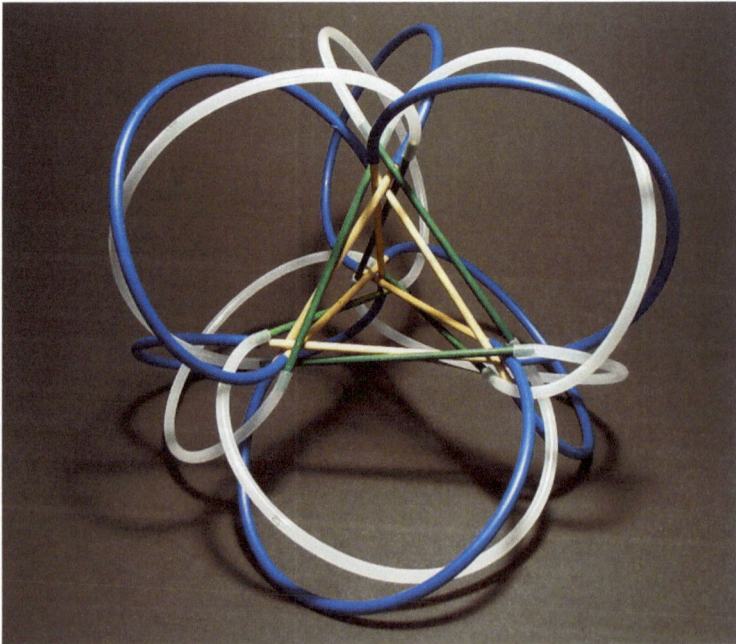

Fig. 2. 55 – Anti-Ambi-Structor (Anti-Ambidextrus Structor) (face view) has both real and anti-matter circuitry, produced by three real matter loops in blue, and three anti-matter loops in white. Shown is a clockwise charge field which produces an anti-matter particle.

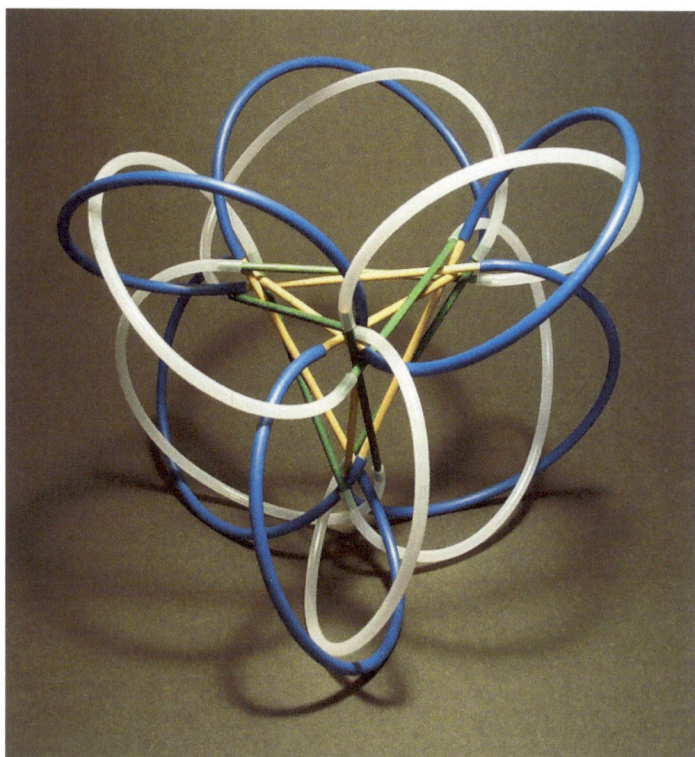

Fig. 2.56 – Anti Ambi-Structor (Ambidextrus Structor) (vortex view) shown-ing white Anti-Matter (AM) loops overtop (cloaking) blue real-matter (RM) loops. This figure is showing both RM and AM loops in there deployed positon. In nature, this would instantly change into Fig. 2.57 whereby the cloaked wave would condense to the nucleus (defined by the yellow sticks) due to EM forces when arranged in this fashion.

Structors model particle structure. Particles all have a nucleus, defined by the sticks portion of the loops in Figures 2.55 & 2.56. In an electron, the particle's nucleus is the mass domain delineated by the tetrahedron, and the extended loops are the electron-neu-trino action field associated with the electron's mass. That is why the word "electron" is attached to the electron-neutrino. The total mass of the electron has to include the little amount of mass (twist) that is in the deployed portion of the loops. Hence the energy mass of the electron is not simply the nucleated portion of the form, but also must include the deployed field portion. If the mass of the electron's nucleus is considered as one, then the total energy of the electron that includes the electron-neutrinos in the charge field is 1.000069809+ .

The *Structor* family of fieldstructures – the form taken by particles

---

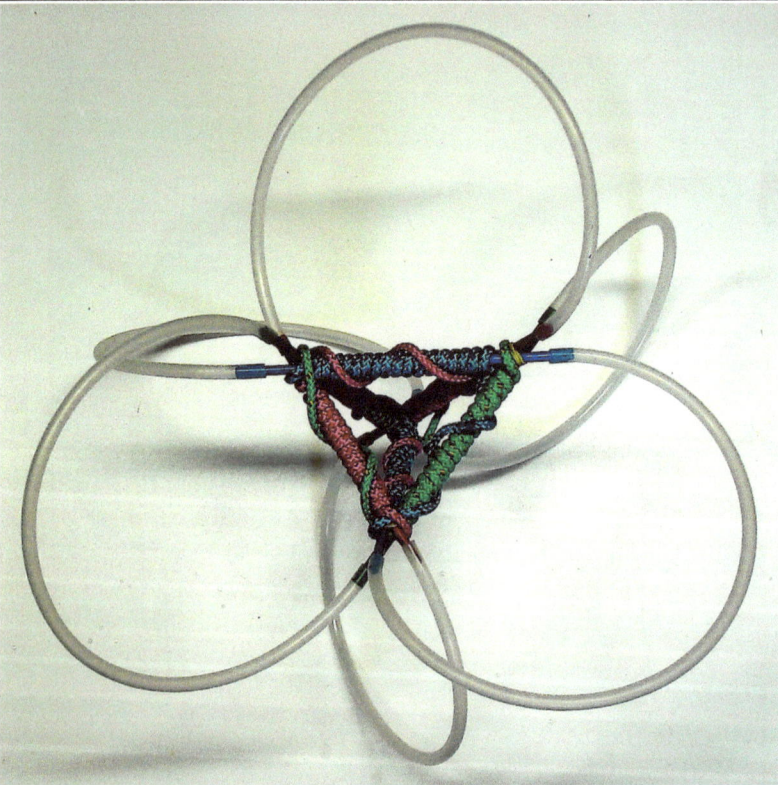

Fig. 2.57 – Particle Platform – A form having a polyhedron nucleus made by loops of action. For example, the electron's structure has three EM wave-sets each having one side of the wave-set's frequency condensed to form and energize the nucleus. As can be seen, the nucleus takes all but 0.06% of the energy (twists) to the nucleus. If the handedness of the model above is reversed, the anti-electron condenses to become an anti-proton and the positron wave becomes the field particle.

A feature to note is that the fieldstructure models show how the three quark loops interact to produce a fermion particle. A wave becomes a particle when the wave condenses. It becomes a field particle with a charge when one wave is deployed and the other is condensed. Waves condense to produce particles because they become space creating events. Waves make particles when the waves form a fieldstructure.

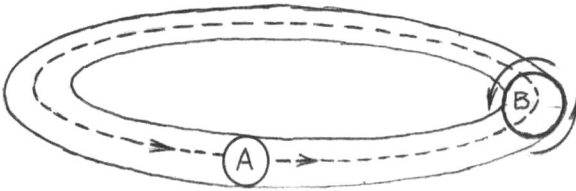

Fig. 2.58 – A loop produces an action event (energy) by completing a 360° rotation (A) about an external locus, which may be circular as show in the figure, or it may be any other shape no matter how convoluted. A loop can also twist around its internal axis (B). Think of (B) as a clock with the hands rotating about the dial keeping time, and the loop (A) as the motion of the clock through space. (A) is the space component and (B) is the time component of space/time. However, many times the hour hand loops around the clock (A), the minute hand rotates around its axis a fixed number of times. In nature, (B) is the square of the loop number (A). Multiplying the number of loops times the number of loops squared is the total rotations in the form and that is its total energy.

A loop has two forms of energy. (1) A string, when made into a loop, gains a unit of energy, because to make a loop from a string without twisting the loop as you make it, the string has to rotates 360 degrees. When the string is twisted 360° around its axis, a unit of energy is added to the string. To ascertain the energy of a loop, each twist is a unit of energy, a quantum. When the twisted string is made into a loop, the twists in the string are locked into the loop. A kinetic energy loop (i.e., EM energy) cannot be destroyed in a permanent sense. If a natural energy ring is cut, it will heal itself and reform into a loop again. In that sense the loop cannot be destroyed. This phenomenon can be empirically demonstrated by cutting a smoke ring and watching it heal itself. This phenomenon is called the **Heal-Cut Mechanism.** It is why FST can state that a line/loop of action cannot be destroyed (Tometry Axiom 2 on page 82). The Heal-Cut Mechanism demonstrates the Conservation of Energy Law in thermodynamics.

In Field Structure Theory, the energy equation is:

$$E = O(\omega^2)$$ wherein O = loops and ω^2 = the number of twists in the loops squared.

Fig. 2.59

As previously shown, a loop action field (Fig. 2.54) produces a polyhedron nucleus. Ironically, Tometry brings back the usefulness of solid geometry to particle physics. It turns out Euclidian solid geometry, with loop circuits added, is crucial to understanding the structure of particles and atoms once the geometry is skewed, knotted and considered topologically.

Any polyhedron, however complex, can be circuited into a fieldstructure using loops.

Because fields are built with loops, and every form of matter is derived from electromagnetic loops, every form in nature has an action field and every action field has a specific field object polyhedral nucleus associated with it. By knowing the polyhedron of a fieldstructure, the mass in loops is known. By knowing the mass loop number, the circuitry and frequency of that loop circuitry is known.

This model supersedes the non-determinacy of the Heisenberg Uncertainty Principle that says position and momentum cannot be known at the same time. In FST, by knowing frequency of a fieldstructure, the polyhedron form of the nucleus can be determined. If the polyhedron is known, the frequency of the form is known. This is because every 3-D fundamental form has a unique frequency that is uniquely linked to a specific polyhedral nuclear form. Thus, the positions where particle interactions are occurring, simultaneously, reveal the momentum of the particle in terms of frequency. The implication is that a particle's position is determined by the interacting waves. Since there are more than one place the loop/waves are interacting in a fieldstructure, there are multiple places the particle can be found. This is the **Multiple Certainty Principle.**

There are four places a particle can be found in the simplest fieldstructure, the Tet-Structor (Fig. 2.54). Atoms, with the electrons whirring around at close to the speed of light (SOL), move in unison processionally so the pattern of vortices becomes seemingly everywhere forming a virtual solid cloud of charge. Without knowing the structure, this solid cloud of charge gave rise to the quantum mechanics having to deal with location in terms of probabilities. There is nothing wrong with quantum mechanics. It just isn't sufficiently deterministic to give a true picture of what is going on with energy at the fundamental level. The value of being able to reproduce the structure of a particle or atom at the macroscale is that the motion of the form can be brought to a stop and studied.

Each polyhedron gives a different circuitry pattern and unique arrangement of vortices. Vortices are where the lines of action interact. These vortices are the places where a particle can be found. A tetrahedron has four places where three loops interact. In a cube there are eight places that waves interact. In an icosahedron there are twenty places where the lines of action interact. Each vortex in a fieldstructure is where a particle can be found inside what quantum physics calls the charge cloud. Since **there is more than one vortex in a 3-D form, the number of places a particle can be found depends on the number of wave/loops in the structure and the frequency of those waves as**

determined by the geometry of the polyhedral nucleus. Hence, the non-determinacy of the Heisenberg Uncertainty Principle can be replaced with the more accurate determinacy of the **Multiple Certainty Principle** (MCP), which states:

The frequency of a fieldstructure establishes a specific and discrete number of places that waves interact and it is only at those places a particle can be found and at no other places.

Stated differently, if the polyhedron nucleus is known, the number of places a particle in the system's charge field can be found is known, and at the same time the energy (frequency) of the system is known.

The following are examples of fieldstructures each have a different polyhedra nucleus, a different number of circuits (loops) and a unique wave frequency. The tetrahedron has been shown previously in Fig. 2.54

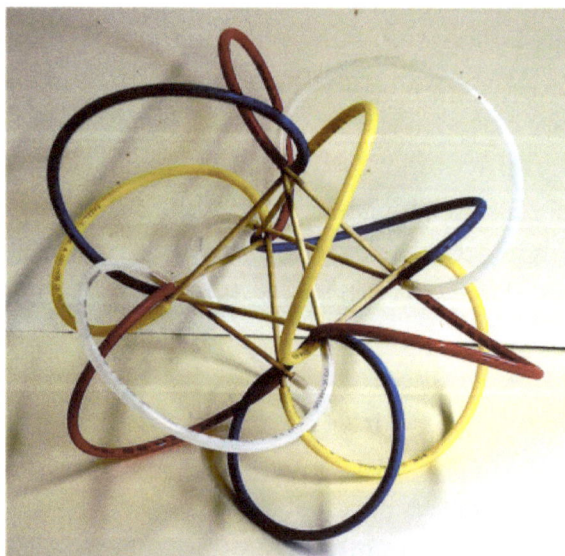

Fig. 2.60 - Octahedron Structor - having a Multiple Certainty of 6, which are the places where the four loop/circuits interact at each vortex. This Octa-Structor shown above has a clockwise spin.

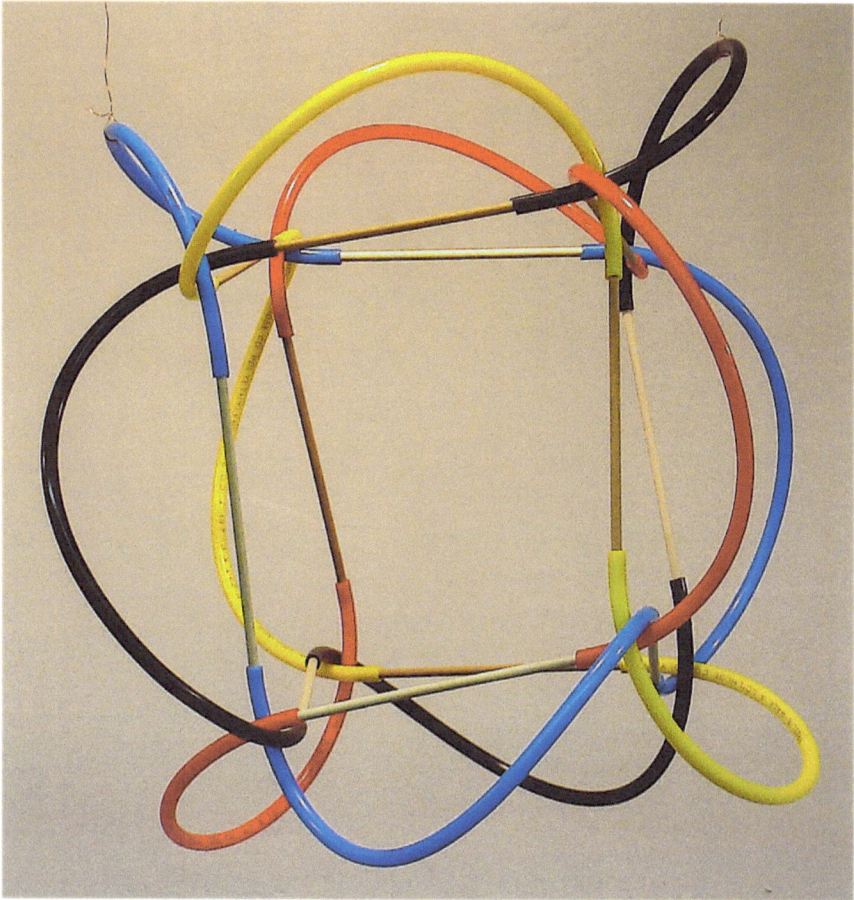

Fig. 2.61 – Hexahedron Structor (cube)- having a Multiple Certainty of 8, which are the places where the four loops/circuits interact at each vortex. This face view of a Hexa-Structor is spinning counter-clockwise at each vortex.

Note how each polyhedron generates its own particular wave frequency. Shown above are the five platonic polyhedra. All polyhedra will have their own unique frequency wave structure. This makes the atomic platform a tesseract with field lines. This makes a SuperStructor a fourth dimensional polyhedron. Frequency of the form is determined by the number of

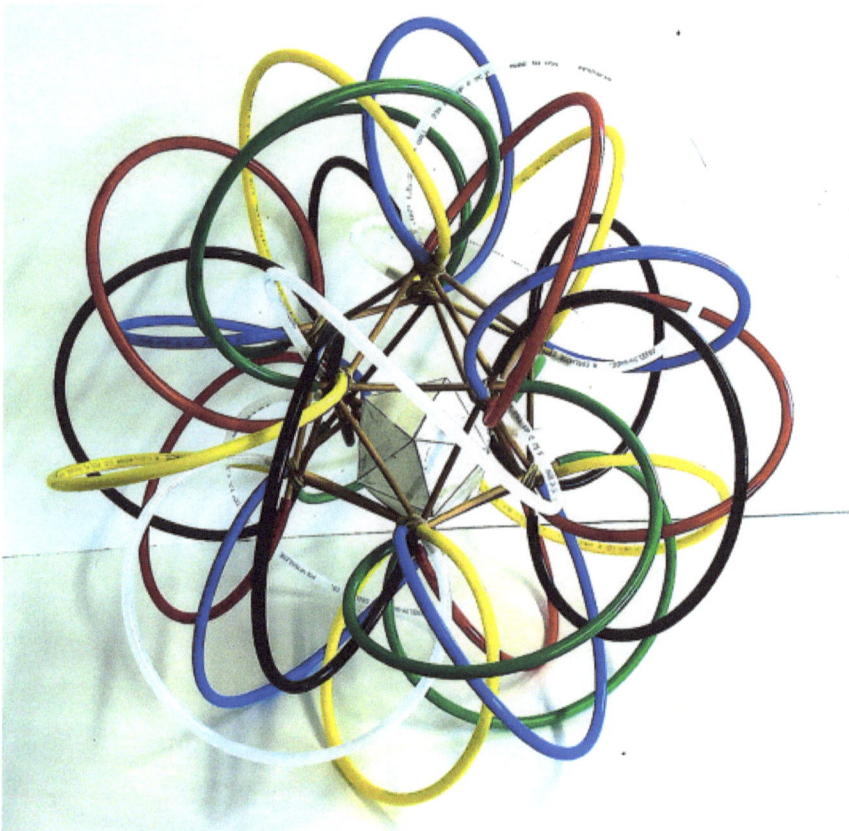

Fig. 2.62 – Icosahedron Structor Fieldstructure - having a Multiple Certainty of 20, which are the vortices of the icosahedron. This face view of an Icosa-Structor spins clockwise. Note that at this level of complexity, the torus shape of the loop/circuits can be seen in the small picture (Fig. 2.63). It is as if five waves are rotating around a center that is not occupied (void of waves). In this photo, the icosahedron nucleus is shown as a solid white cardboard form which was used as a scaffold to build the structure.

interacting loops and the particular polyhedra nucleus of the form. In solid geometry, all 3-D forms are polyhedra. Any 3-D polyhedron can be created with loops be they regular (Platonic), semi-regular (Archimedean), or irregular polyhedra. The loops are to be understood as the action field of the form. Structors in

Fig. 2.63 – a single loop circuit of an icosahedron. Note how the loop describes a torus form. Each loop in Fig. 2.63 forms a torus in a separate plane all of which interact to form a spherical field around the nucleus as seen in Fig. 2.62.

Figures 2.54, 2.60, 2.61, 2.62, 2.64 model the circuitry of action of the Platonic solids. In nature, the five noble gas elements have the geometry of the Platonic polyhedra. Because the Platonic solids are everywhere symmetrical and perfectly balanced, in nature they are the noble gases. Noble gases are symmetrical atoms and thus do not interact with other atoms, or even with each other.

Fig. 2.64 - Dodecahedron Structor Fieldstructure - having a Multiple Certainty of 20, which are the places where the five loop/circuits interact at each vortex. This vortex view of a Dodeca-Structor spins clockwise. Note how each loop/circuit rotates around the nuclear polyhedra two times whereas the icosahedron which also has 20 vortices only goes once around the nucleus. While the icosahedron and the dodecahedron are duals of each other, both have five loops, but have different frequencies.

27 - The *SuperStructor* – The Fieldstructure Form Taken by Atoms

SuperStructors are circuits of EM energy that link a smaller internal polyhedra with high energy to a larger external of polyhedra at a far lower energy. The SuperStructor models the circuitry of action in atomic structure (Fig. 2.65 & 2.66).

In SuperStructors, an external field polyhedron is circuited to an internal field object polyhedron. This describes an atom. Atoms have field polyhedra (the electron shells). The charge field polyhedron (yellow sticks) seen in the model below are what the Standard Model has been calling the electron cloud. The atomic cloud it turns out has a definitive geometry as does the nucleus.

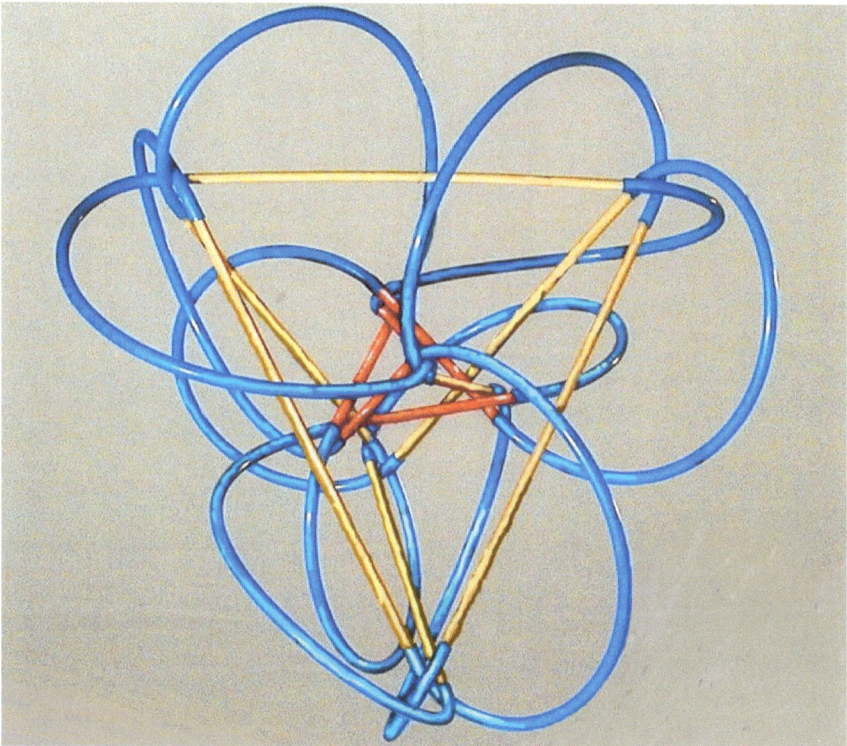

Fig. 2.65 – Tetrahedron SuperStructor – vortex view. Three loops circuiting a small nuclear tetrahedron in red is inside the large yellow tetrahedron that forms the polyhedron charge field (yellow sticks). The small inside nucleus in red is spinning clockwise. The large outside tetrahedron in yellow is spinning counter-clockwise. These spins model how the proton nucleus spins clockwise and thus has a positive charge, while the electron in the charge field has a counter-clockwise spin and negative charge.

Both the nucleus and the charge field have polyhedron geometry. Shown in Fig. 2.65 is an atom with one shell, i.e. the hydrogen atom. There can be up to seven shells, each a polyhedron inside a polyhedron in layers.

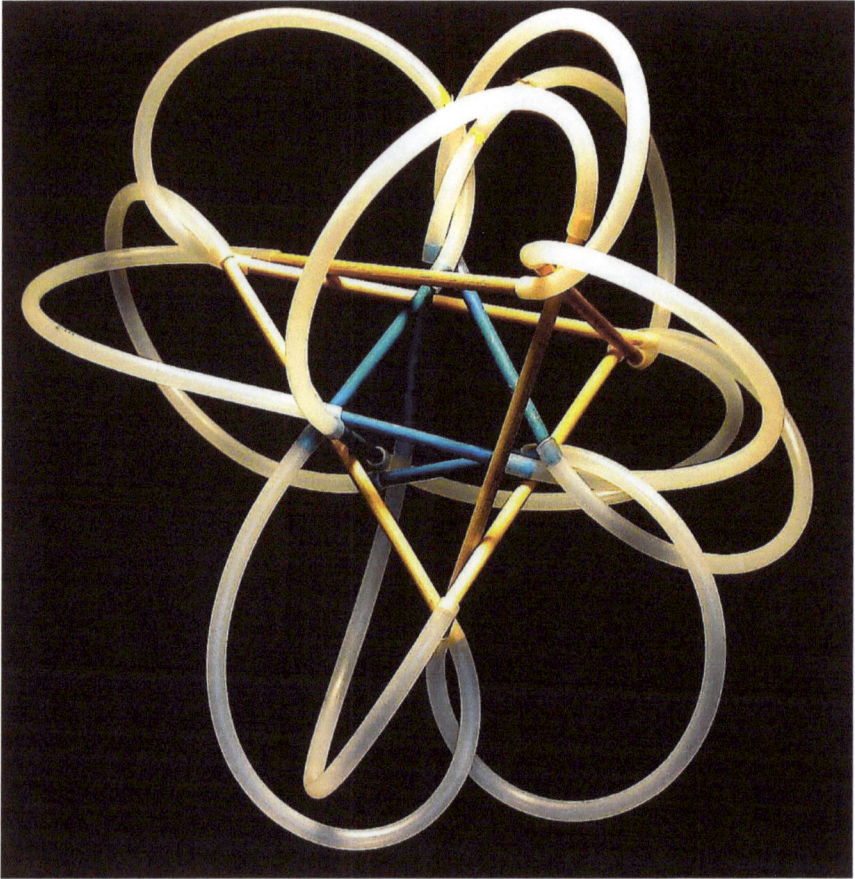

Fig. 2.66 - Tetrahedron SuperStructor – face view

In an atom, the field is populated by particles at the places in the field that have three or more loop circuits interacting. These places of interaction are where the electron can be found. Because there are three loops interacting in the field at the vortices of the field, the particles at those places are fermions. They are a fractal rendition of the nucleus. Add all the energy of these fractal electrons and the node count (energy) will be the same as that of the nucleus. The interactions have a handedness (chirality). The field has a charge that can rotate either clockwise or counter-clockwise. In physics, a clockwise rotation is a positive charge and a counter-clockwise rotation is a negative charge.

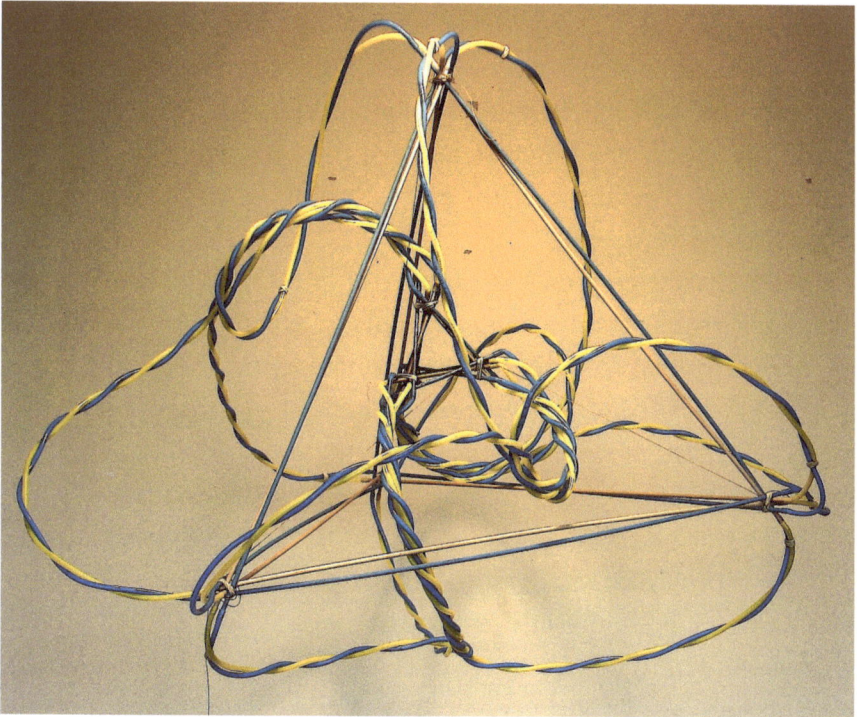

Fig. 2.67 – SuperStructor showing how the inner and outer tetrahedron can be at a considerable distance from each other as is the case in atoms where the nucleus is 100,000 times larger than the volume of the nucleus. Note how the chiral EM lines of action twist numerous times between where it leaves the nucleus and joins the external tetrahedron. The large sticks are only to show the polyhedron that the form has, although in nature the polyhedra inside and outside are curvilinear. In more complex polyhedra, the toroid shape of the circuits become quite evident. See Fig. 2.63 for an example of how the toroid circuitry of the charge field becomes increasingly more evident the more complex the polyhedron becomes.

Shown in Fig. 2.65 and 2.66 is the circuitry of action in an atom having one electron in its first shell, i.e., hydrogen. The double wave-set structure seen in Fig. 2.55 and 2.56 is not shown in Fig. 2.65 and 2.66. Only the one side of the wave-set is shown in Fig. 2.65 & 2.66 so that the configuration can be seen clearly, which is to say the electron wave is shown and the positron wave is not shown. In stable natural elements there can be up to seven shells circuited to the nucleus having up to eighty-one protons. Eighty-one stable elements and eleven unstable elements are found in nature's 92 natural elements. An eighth shell is not found in nature. The kinetic energy of the electron is too high for the EM attractive force of the nuclear protons to hold the electrons in an eighth shell.

Atoms arrange the electron circuitry according to the energy of the electrons. The number of shells in an atom is determined by the complexity of the nuclear polyhedron. That complexity is governed by how many protons there are in the nucleus. An interesting structural question is, "Why do the shells have varying numbers of electron circuits per shell?" This requires further study of the relationship between the nuclear polyhedron, circuitry patterns, and separation distance between electrons among other factors. Presently, the energy of the shells is known, but not why shells arrange themselves as they do. What can be explained is why there are only seven shells possible. Further modelling is needed to answer these questions.

28 - The *Mala* – The Fieldstructure Form Taken by Molecules

Loops from one atom's charge field can connect to another atom's field by sharing an orbital particle called a *valence electron*. In atoms, the valence electron performs the connectivity function by circuiting the atoms into a molecule.

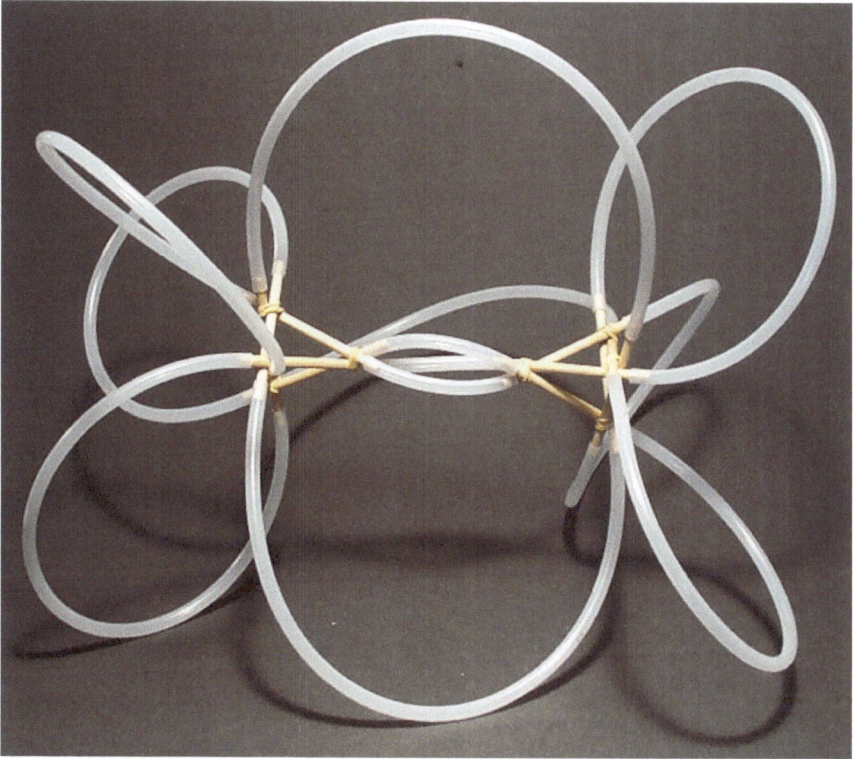

Fig. 2.68 – A valence loop circuiting two atom nuclei (the stick tetrahedrons). Not shown are the non-valence electron's circuits that always remain with each atom. Valence electrons are the electrons that can be shared with other atoms to form molecules. Illustrated are three valence loops circuiting the molecule.

In the Mala Structor shown in Fig. 2.68, a valence electron orbits between six atomic nuclei shown as the stick tetrahedrons. In a molecule, a single electron can do the work of two or more electrons. This shows structurally how valence electrons efficiently join atoms together, doing more with less. Fewer electrons are needed when atoms are conjoined. Any combination of polyhedron can be circuited in this manner as long as the electron shells are not filled. The non-valence electrons remain bound to their respective atoms and do not participate directly in the

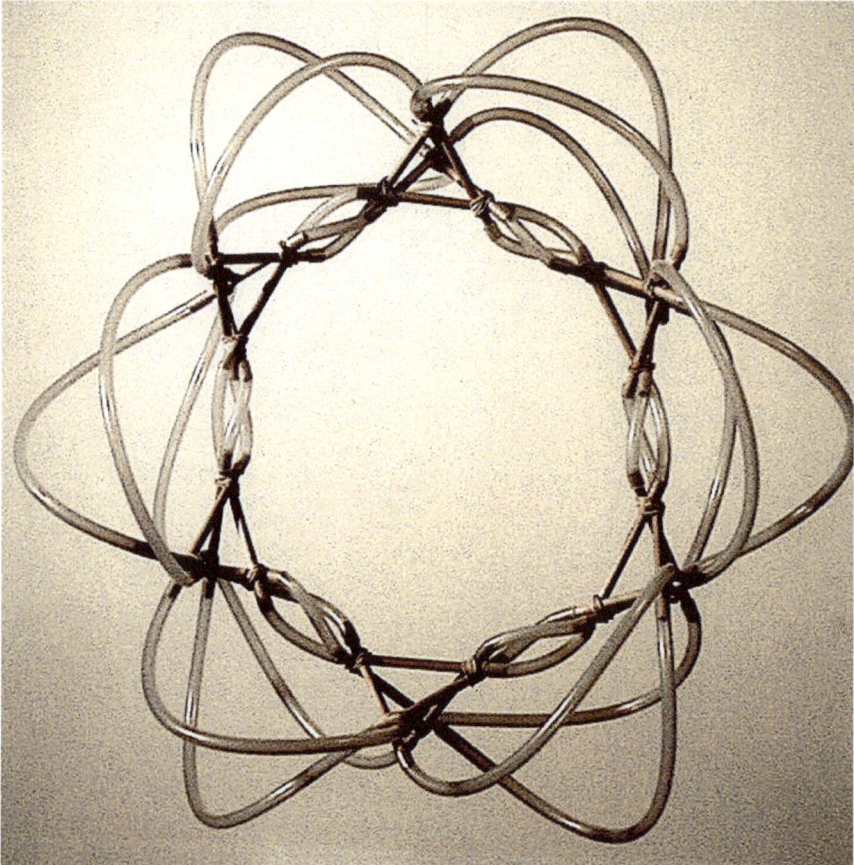

Fig. 2.69 – A single valence loop circuits six nuclei (atoms).

molecular circuitry. In particles, atoms and molecular structures, loops reveal the shortest path for action to take in order to circuit atoms together and provide a balanced structure in equilibrium. Equilibrium is the overarching imperative.

29 - Fields

Traditionally fields have been diagrammed, but not explained in any causal structural sense prior to the discovery of fieldstructures. Scalar and tensor fields are diagrammatic descriptions of certain particle attributes, but they are not structural in the sense that Field Structure Theory would define. Fields in physics are maps of the energy potential the way a topo map of a mountain range shows in concentric bands the elevation of the ground. It tells us nothing of why or how the ground is elevated. It tells us what the energy is of a place and that is all. Such mathematically descriptive maps are not explaining what is going on structurally.

Loops create force fields the strength of which is determined by the degree to which the loops are condensed and twisted. This makes it possible to build a working model of a static field that behaves in the macroscale the same way the kinetic field does at the atomic scale if the kinetic field was frozen in time.

As a field gets smaller in volume, its energy increases, due to the condensing of the wave into a high-frequency more compressed form. Hence the binding energies of Higgs particles (if they in fact exist) requires trillions of electron volts (TeV) to exists. While a Higgs particle would store trillions of electron volts, nucleons require billions of electron volts and electrons require millions of electron volts. Molecules require thousands of electron volts and cells require hundreds, while humans cannot withstand but a the few electron volts before structural changes begin to destabilize the body/mind complex. As each platform of structure decreases in bonding energy, the space in which the energy resides proportionally increases.

Spatially, matter does the opposite of what energy does. As matter becomes less energetic, its spatial extension becomes larger. A proton requires a spatial domain that is 10^{-15} meters. The electron lives in the atomic cloud that is 10^{-12} meters, 100,000 times bigger than the nucleus. A molecule lives in fields between 10^{-10} and 10^{-8} meters. Cells begin in size at 10^{-7} meters and top out around 10^{-3} meters (the size of the human egg). The universe is a hierarchy of fields going from large to small (universe to particle) or if you prefer, from small to large (particle to universe). Each field is connected by looping to all other fields. The universe is loopy. Theoretical science strives to understand how practical transformations can occur between fields giving mankind the ability to manipulate reality (for good or bad) to suit their designs. It is in those transitional stages between the *start-state* and the *end-state* that science will be able to focus once the field structure of the form is understood. From the FST perspective, it is these transitional states that are revealed by the circuitry schemes of action loops.

A field distributes energy uniformly to achieve balance, by enabling stress in the system to be shared equally to all parts of the system. It is generally recognized that mass is a field force that is sequestered to a proscribed domain. Mass is a condensed field. Hence, it can be said that only fields exist. All fields are connected to each other, including mass fields. Given this fact, physics has seen the necessity to formulate a Unified Field Theory.

Fields have two states:
    (1) those that are **deployed**.
    (2) those that are **condensed**.

Kinetic energy is the flux between those two states, i.e., between deploying or condensing loops. All phenomena are energy fields. The four known field forces are gravity, electromagnetic, and the weak and strong nuclear forces are degrees of loop condensation. What is missing in physics is the Plenum field, the mother of all fields. By implication, finite closed fields are at the same time connected to the Plenum. FST considers the Plenum to be a field, the mother of all fields and the stuff of creation.

The four condensed fields of gravity, electromagnetism, the strong and weak fields are the product of the Plenum loops being twisted and knotted so that they can achieve a stable three-dimensional

interaction. Adding interacting loops creates the hierarchies of form, beginning with EME, and ending in a being that can comprehend the Totality.

The above statement is a bow to honor the descriptions modern physics has given us about field. While FST regards fields differently from the Standard Model, it is physics that got us to the point a Field Structure Theory can be understood and confirmed. Physics in the past has seen the four fields as distinct. FST sees these fields as states of loop condensement and deployment. All fields are loops. What has been missing is an understanding of how each field is structurally related to each other.

Energy is either centrifugal (deploying) or centripetal (condensing). A mass is a fields that distribute the centrifugal and centripetal forces equally within a domain to achieve balance and stability. Gravity is the tensional effect of having the Plenum condensed to form matter. This condensing results in compacting the Plenum. Compressing the Plenum causes gravity. The Plenum constantly attempts to remove the tension created by mass by trying to pull mass apart in an effort to break the mass knot. Curiously, to pull a mass apart means the mass has to be compacted. This is because mass is a knot. It is the function of a galaxy to pull matter apart and the function of a star to pull mass together.

It may seem counterintuitive to suggest gravity is the attempt by the Plenum to pull mass apart, in view of the fact we experience gravity as an attempt to squeeze matter together. Gravity has been seen as concentrating mass, not pulling it apart. This can only be understood by studying the forces at work in a fieldstructure.

However, from the Plenum's point of view, the Plenum has been inverted by the knotting process that occurs with the formation of mass. Pulling in causes heat and pressure. It is heat and pressure that can break the knot holding mass together. Nature's mechanism for that is found at the core of a galaxy.

Squeezing hydrogen atoms together makes helium with the release EM energy. The tension in the Plenum (gravity) is released by destroying mass. In separate processes, Plenum both creates and destroys mass.[6]

The form that prevents the condensed energy from redeploying back to the Plenum is the FieldStructure form. This holding together phenomenon is called syntropy . It is the opposite force of entropy. The Plenum without mass is without tension. With no tension there is no energy, no gravity. Fieldstructuring is nature's way of sequestering energy so that a stable universe can be created that will endure in time. Nature invented, so to speak, tension and compression so we could have a universe as we know it.

The various fields of condensation, the Aum, Ark, Bark, Lark, Spark and Quark fields (the **A-Q Fields** for short) are each a degree of field condensement. The smaller the volume of a field domain, the stronger the binding force of the field. The looping of the nodes become increasingly compressed so that the number of nodes per unit of length increases as the field compacts. As the loops compress, the energy of the field increases. This increases the wave's frequency. To break the field apart, requires more energy than the particle or atom etc., can accommodate.

This leads to the question: "What condenses the loops and what keeps them condensed?"

The answer is "tension" in the Plenum.

If so, what is capable of applying tension to the Plenum since by definition there is nothing outside of the Plenum capable of acting upon it?

The answer is *consciousness*.

Is consciousness outside the Plenum?

NO.

---

6   There is no contradiction in something able to create and destroy when Totality is at issue. After all, human beings and all creation is doing both all the time; planting a garden and then eating it.

The question asks, is a string outside (separate from) the knot? It can't be.

Consciousness is the nature of the Plenum.[7]

You know you are entering the world of metaphysics if the issue being considered has no causal agent. Metaphysics is the world that has no dialectics at play and yet is the substratum form which dialectics comes into being.

Hopefully in this century, science will recognize the role of metaphysics, consciousness and the Plenum in the formation of reality, something the yogis and yoginis have known for millenniums.

Having now sketched out the structure of particles, atoms and molecules, showing how each builds upon the other with loops, we can quantize this hierarchy of loops to obtain the mass and energy particle values. This we have done in Chapter One of this book.

Let me close with a charming legend told to children by Field Structure Theory elders.

The Mother Field of the universe, sitting on her throne, had a beautiful ring on her finger. She had two sons and they both asked her for her ring. Not willing to choose one over the other she gave them a challenge. The challenge was to circumambulate the entire universe and present themselves back before her. The first to arrive would get the ring. At that the big brother smiled. It was common knowledge that he was the epidemy of strength, speed, endurance and power. His physical prowess could not be challenged.

The younger son was overweight, slow, and had physical handicaps that made him no match physically to his big brother. The younger son, however, was known for his intelligence.

---

7   To get the full description of how this works can be found in numerous places in the Vedas.

The race began and the big son in an instant was out of sight as he sped away at the speed of light. The younger son with great effort picked himself up and shuffled around the Mother Field's throne arriving back in front of her seconds before the big brother returned after circumscribing the entire universe. The younger son was given the ring. He correctly perceived that the entire universe is embodied in the Mother Field herself. The Plenum is such a Mother of all Fields.

Field Structure Institute
Don Briddell
www.FieldStructureInstitute.org
donbriddell@fieldstructureinstitute.org

TERMS AND DEFINITIONS

Field Structure Theory and Tometry may have new meanings attached to old terms. Where definitions are well known those meanings can be googled for brevity. The special character (*) denotes that a new term has been introduced to explain a new concept.

Action Field*: The loopage in a fieldstructure that is outside the polyhedron nucleus. In Nature, the charge field generates the field object; formerly called the "electron cloud" in Standard Physics. Action fields have charge because one side of the EM wave-set has condensed while the other side of the EM wave-set is deployed.

Anti-Matter (AM): The same as real-matter except having the opposite handedness and spin. Anti-matter is the unseen, unseeable, inverse of matter responsible for waves and structure. Indispensable to the formation of reality; equally abundant , equally present and entangled with real-matter which cloak it from view.

Ark*: Electromagnetic energy, the first loop to condense from the Plenum, having the largest volume, obeys boson statistics, two chiral forms – right and left-handed.

Aum*: The loop family from which all form is derived, the Plenum. Pervades the Universe, contains all energy. The mother of fields.

Axioms: The assumptions that determine a proposition. Can only be deemed correct within their ability to formulate a meaningful conclusion.

Background Independent: The idea that all matter is separate from the underlying Plenum (ether) as a bead is separate from the string on which it is strung. The idea that something can be in space, but not attached to space.

Background Dependent: The idea that all matter is a field of energy derived and inseparable from the Plenum, as a knot is inseparable from the looped string on which it is tied.

Boson: Energy in a wave form. Electromagnetic energy. Phenomenon that obeys Einstein-Bose Statistics.

WB-FS*: (Wysocki/Briddell – Fieldstructure) is a fieldstructure defining a nucleus and charge field. Having a fermion wave structure. The structure of three loops (or more) that Interact forming a 3-D knot of stable energy which creates a 3-D space. Obeys fermion statistics.

Clinton – Fieldstructure*: The deployed form of a fieldstructure. A spherical field having no nucleus, obeys boson statistics, has energy that is evenly distributed in the action field.

Charge Field: That area of a fieldstructure that extends from the nucleus to outer limits of a particle or atom. In real-matter, it is the domain where the electron's orbital circuit is outside the nucleus. The field has a charge because the wave-set has had one side of the wave condensed leaving the other side deployed. That separation creates a force (charge). The force is in the form of the condensed loop/wave trying to deploy into the charge field loop/wave, the strength of which is dependent on the number of twists in the loop, i.e., the waves frequency.

Condensement*: Compressing the nodes of a wave into a smaller volume thereby increasing the energy (frequency) of the wave. The number of nodes on the wave does not change, but the distance between nodes decreases when condensed.

Deployment: Expanding the nodes of a wave into a larger volume thereby decreasing the energy (frequency) of the wave. The number of nodes on the wave do not change, but the distance between nodes increases when deployed.

Design Science: Combines design and science to find a bridge between art/humanities and science. Buckminster Fuller is considered its founder. Design scientists consider themselves to be the charioteers holding the reins of two horses, that of art and science, insisting they work together.

Determinism: The idea that reality is knowable and can be determined even at the atomic and particle scale, limited only by the acuity of the investigator. In FST, determinism is achieved with the Multiple Certainty Principle.

Downerate: Making fractals smaller.

Electron:, The electron is the first particle in the Electron Sequence, having three Spark loops. In Standard Physics, an electron is a lepton with 0.511 MeV energy, a number that does not include the mass of its electron-neutrino charge field particle. The mass of the electron is 1.000083154929194, a measurement that includes three electron-neutrino mass units for a total loop value of **0.000027718309765.**

Electron Sequence*: The hierarchy of particles found as iterations of the Sierpinski Triangle Fractal that begins with the electron which has three loops (the topmost triangle) and end with the lambda hyperon triangle having 2187 loops. Any particle having less than 2187 electron masses is in this sequence. The electron sequence includes all 1st generation particles, excepting the neutrinos that have their own sequence, the Neutrino Sequence.

EM: electromagnetic

EME: electromagnetic energy

Energy: Manifests when one side of the chiral electromagnetic wave/loop-set condenses or deploys. Condensing the waves compacts energy to produce mass. Deploying the waves lessens mass and produces radiant energy.

Fermion: Fermions in Standard Physics are particles having an energetic rest state mass that define a local spatial domain with a polyhedron nucleus. Fermions obey Fermi/Dirac statistics.

Field: All form and structure are field forms. The degree of condensement determines the mass of the field. Fields can be condensed or deployed or both. A function of the field is to provide equilibrium. All fields have a limited spatial extension. Fields are condensed Plenum. Only the Plenum can be without condensement.

Field Object*: A knot on a string is analogous to a mass in a field. A field object refers to a nucleus, be it a particle's nucleus or an atom's nucleus.

Field Structure Theory (FST)*: The application of Tometry to describe and explain the natural world. A 3-D interaction of three or more loops that are knotted together. Can be condensed into either the Wysocki/Briddell-Fieldstructure (WB-FS) or deployed into the Clinton-Fieldstructure (C-FS). In nature, real and anti-matter loops are entwined in a fieldstructure. The energy of a fieldstructure is determined by a count of the loops and the number of twists in those loops. $E = L(l^2)$ Energy equals loops times loops squared.

Fieldstructure*: The interaction of three or more loops that are knotted together. Can be condensed into either the Wysocki/Briddell-Fieldstructure (BW-FS) or deployed into the Clinton-Fieldstructure (C-FS). Fieldstructures are the product of interacting EM loop/waves.

First-Generation Particles in Standard Physics: In the Standard Model (SM), the first generation of particles begins with the neutrino and ends with the neutron particle. In Standard Physics, the particles are arranged with no internal hierarchy other than mass. In FST, particles are arranged by their loop or twist counts.

First Generation Particles in Field Structure Theory*: In FST, the 1st generation of particles begins with the electron-neutrino and ends with the Tau-neutrino. The 2nd generation begins with the electron and ends with the lambda hyperon. This sequence arrangement is governed by the Sierpinski Triangle Fractal (STF). Each generation has seven iterations in the STF.

Frequency (in Tometry)*: The number of nodes on a wave. Frequency is caused by the helical interaction between real and anti-waves moving in opposite directions forming a wave-set. All waves come in wave-sets. In Tometry, the frequency is determined within the context of an individual loop/wave. It is *not* a measure of a wave-train which is a count of the number of waves that pass a point in a period of time, such as measured by a linear oscilloscope.

Heal-Cut Mechanism*: The ability of a wave-set to reform into a loop after being cut. The key mechanism in wave transformation from one energy state to another. The mechanics of the Conservation of Energy Principle. A stone dropped into a pond creates ripples. Each ripple is a chiral wave having clockwise and counter-clockwise energy rotating around the wave. That is why a wave reforms on passing a pole. The two circulating chiral waves close back to re-establish the wave.

Knot: A form that action can take to confine energy to a locality. A knot is a unit of energy, a quantum. A 3-D knot creates space, whereas a lower dimensional knot such as discussed in topology, lives in space but doesn't create space as does a true 3-D knot in Tometry.

Lark*: The loop family of neutrinos

Loop Families*: Aum, Quark, Spark, Lark, and Ark are different degrees of loop condensation with Aum loops being the most condensed and Ark loops the least condensed.

Loop-sets / Wave-sets*: Terms are interchangeable. The word "loop-sets" used in Tometry and "wave-sets" are used in structural physics. In Tometry, a loop-set is entangled right and left-handed loops sharing a common axis and spatial domain. In Nature, a wave-set is the Tometry form of electro-magnetic energy.

Loops & Looping*: The postulate that energy in all forms is a closed system of action.

Macroscale: As used in FST, it refers to those things 10^{-5} to 10^{+5} meters in size. Macroscale forms are the things at the human scale as opposed to microscale and cosmic scale things.

Mala: In Tometry, a linkage of SuperStructors. In Nature, Malas are molecules, B-MA-FS (Briddell-Mala-FieldStructure).

Mass (in FST): Energy sequestered to a spatial locality having a rest mass below which the form does not lose energy. Localized energy. Energy that obeys fermion statistics. All mass forms are fieldstructures. Mass is a count of loops times the number of times the loop squared is rotated about its axis (twists). Mass equals Loops times loops squared.

Microscale: All things smaller than 10^{-8} meters beginning with molecules and extending smaller in scale to the lengths, volume, or mass of a Planck's particle.

Multiple Certainty Principle*: Matter has a specific number of places it can be within a quantum system according to the frequency of the loop/waves, the number of loops in the system, and the particular polyhedron of the nucleus. The principle returns determinism to physics, replacing the indeterminism of the Uncertainty Principle with a more accurate multiple certainty description of how energy at the fundamental scale behaves.

Neutrino: A single electron-neutrino has a mass **0.00000028728015** MeV. Neutrinos are the field particle of the electron. There are three energy forms of the neutrino. The form of the neutrino is composed of three Lark loops.

Neutrino – sequence*: Begins with the electron-neutrino and ends with the tau-neutrino and has seven iterations. The neutrino-sequence has STF hierarchy. Three tau-neutrino equals the ground state energy of electron.

Neutron: A neutron is a proton that has an electron along with the electron's electron-neutrino condensed to the proton's charge field. The charge of the proton is neutralized by the electron transforming it into a neutron. The electron and electron-neutrino cycle between two protons. The condensed electron becomes a gluon and changes from a fermion to a boson. This means a boson is a condensed fermion.

Occam's Razor: The idea that the simplest solution is most likely the right one.

Particle field*: That portion of a particle's fieldstructure that defines a polyhedron; understood in physics as the nucleus of a particle. The particle field is the condensed EM wave/loops of a fieldstructure.

Particle Hierarchy*: The natural hierarchy of particles that are iterations and combinations of iterations of loop assemblages of the Sierpinski Triangle Fractal.

Platform*: The idea that particles form platforms of structure at various energies as dictated by the Sierpinski Triangle Fractal.

Plenum*: The substratum loop matrix from which all energy and matter are derived whereby the loop loops and interact to create form; the Mother of all fields. A field that connects all other fields of lesser size and power. Not limited by space and time until condensed into lesser fields. Action travels instantaneously.

Positron wave / Positron loop*: The left-handed clockwise rotating electromagnetic wave/loop that is entangled with the right-handed electron wave to form an electromagnetic wave.

Probabilities: Nature understood in terms of generalities. The bases of quantum mechanics.

Proton: A positron wave/loop condensed 100,000 times forms a proton. Spin rotation is left-handed clockwise. Mass measured in electron masses that includes the mass of the electron's field particle the neutrino is 1836.1257+.

Quark: Quarks are condensed Plenum loops to the volume of a proton. Three quarks form a nucleon. Spark loops (leptons) condense to become quark loops. Quark loops decay directly back to the Plenum as gamma rays. A quark only takes on particle attributes when three quarks interact at nucleon dimensions.

Real Matter (RM): As real-matter beings, we experience real-matter empirically; anti-matter cannot be experienced empirically. Real-matter and anti-matter entwined in the same quantum system are compatible. RM and AM (anti-matter) when not in the same quantum system mutually destroy the mass structure of each other and become radiant gamma energy. When a part of the same quantum system, the waves do not react to each other. They interact.

Sierpinski Triangle Fractal: Discovered by the Polish mathematician Waclaw Sierpinski. This form has proven to reveal the architecture of matter when each line is considered to be a loop and iterations of the fractal prove to reveal the mass of particles.

SP: Standard Physics

Spark*: Leptons are the Spark Loop Family of form. Electrons are in this family.

Spin: Rotational handedness in particles and atoms. Refers to three-dimensional rotation, not two- or one-dimensional spin. Planets orbit in a two-dimensional plane. Particles and atoms spin in three-dimensions. A basic attribute of particles and atoms.

Standard Model: The model used by mainstream physics to account for mass and energy. A description, but not an explanation.

Structor*: The structural form of the particle platform.

Structural Physics*: The mechanics of how action forms fields using loop and twist analytics.

Structural Platform*: A form of matter that has stability and duration in time.

Structuralist*: A person who models nature with natural forms and structures.

SuperStructor*: Tometry's name for the atomic platform of structure, B-SS-FS (Briddell-SuperStructor-FieldStructure).

Tometry*: A structural analytic using 3-D Loop Topology Fractal Skew Geometry.

Twisting*: The idea that loops can and do rotate 360-degree both around a locus and around their own axis. Each rotation is one quantum unit of energy. It takes a full twist to make a loop in nature.

Uperating*: Making a fractal larger in size and complexity.

Wave-set*: Electromagnetic waves come in a set that combines a right-handed counter-clockwise with a left-handed clockwise rotating loop while sharing the same domain without interference. A wave-set constitutes a quantum system.

BIBLIOGRAPHY

1. **Wholeness and the Implicate Order**, David Bohm, Published by Routledge Inc., NY, NY © 1980/1981/1995, ISBN 0-415-11966-9

2. **The Trouble with Physics,** Lee Smolin, Houghton Mifflin Co., NY, NY, © 2006 ISBN # 13-978-0-618-55105-7

3. **Superstrings - A Theory of Everything?** (note question mark), P.C.EN. Davies and J. Brown, Cambridge University Press, © 1988, ISBN 0 521 43775

4. **Superstrings and the Search for The Theory of Everything,** F. David Peat, Contemporary Book, © 1988, ISBN # 0-8092-4257-5

5. **Six Easy Pieces,** Richard Feynman, published by California Institute of Technology, © 1995, 1989, 1963, ISBN 0-201-40955-0 & 0-201-40825-2

6. **Particle Physics,** Frank Close, published by Oxford University Press, © 2004, ISBN 978-0-19-280434-1

7. **Particle Physics**, Brian R. Martin, published by One World Publications, Inc., © 2011, ISBN 978-1-85168-786-2 & 978-1-78074-039-4

8. **Matter and Antimatter,** Maurice Duquesne, published by Harper & Brothers, NY, © 1960 Hutchinson & Co. Ltd.

9. **Mathematics and the Imagination**, Edward Kastner and James Newman, Simon and Schuster, NY, © 1940

10. **The World of Elemental Particles**, Kenneth EN. Ford, Blaisdell Publishing Co., © 1963, Library of Congress Catalog # 63-8923

11. **The Particle Garden**, Gordon Kane, Perseus Publishing, Helix Books, © 1995, ISBN # 0-201-40780-9

12. **The Structure of Matter**, M. Karapetyants & S. Drakin, Mir Publishers – Moscow, © 1994

13. **Quantum Mechanics**, George Yankovsky, MIR Publishers, Moscow, © 1965

14. **The Neutrino**, Frank Close, Oxford Press, England, © 2010 ISBN 978-0-19-957459-9

15. **A New Kind of Science**, Stephen Wolfram, Wolfram Media, Inc., © 2002 ISBN 1-57955-008-8

16. **Electromagnetism**, John C. Slater & Nathaniel EN. Frank, Dover Publications, © 1947 Library of Congress Number 486-62263-

17. **Fields of Color,** Rodney A. Brooks, Epsilon Publishers, Sedona, AZ, © 2010, ISBN 9780473179762

18. **Connections,** Jay Kappraff, McGraw-Hill, Inc., © 1991, ISBN 0-07-034250-4

19. **Reality Is Not What It Seems,** Carlo Rovelli, Riverhead Books, ©2017 USBN 9780735213937

I would like to thank the Internet and
Google for being a valuable reference source.

Special thanks to Joel Pitney, Laura Hartzell-Pitney and Jessica
Hill with LaunchMyBook.com for their work in putting this
book together.

Om Tat Sat

www.ingramcontent.com/pod-product-compliance
Lightning Source LLC
Chambersburg PA
CBHW041313210326
41599CB00008B/261